SpringerBriefs in Energy

Energy Analysis

Series Editor

Charles A.S. Hall

For further volumes:
http://www.springer.com/series/10041

Graham Palmer

Energy in Australia

Peak Oil, Solar Power, and Asia's Economic Growth

 Springer

Graham Palmer
Paltech Corporation
Knoxfield, VIC
Australia

ISSN 2191-7876
ISBN 978-3-319-02939-9   ISBN 978-3-319-02940-5   (eBook)
DOI 10.1007/978-3-319-02940-5
Springer Cham Heidelberg New York Dordrecht London

Library of Congress Control Number: 2013951764

Printed on acid-free paper

Springer is part of Springer Science+Business Media (www.springer.com)

Preface

A visit to Southeast Asia provides the opportunity to witness first hand the early stages of economic development in emerging economies. Evolving from traditional agrarian societies to societies with greater diversity, Vietnam, Cambodia, Laos, and others are slowly building the essential infrastructure and services that the developed world takes for granted; roads, sewage, electricity, clean water, hospitals, and schools.

A comparison of traditional Cambodian rice farming with modern Australian farm practice perhaps reveals two of the primary differences between traditional agrarian societies and the developed world; the availability of capital, and cheap plentiful energy. In Cambodia, rice is still mostly harvested by hand with a sickle or knife. Each hectare requires about 40–80 person-hours to harvest; an Australian farmer will harvest a hectare in about 15 minutes. The over 200-fold labor productivity superiority is not surprising when one considers that the Australian farmer has access to 500 horsepower of power in a modern harvester, costing a 1,000 times the average annual Cambodian income.

Whereas agriculture makes up around 70 % of the Cambodian labor force, it represents a little over 2 % of Australia's national economy, freeing up the rest of Australian society to engage in the richness and diversity of modern advanced society, all underpinned by the availability of cheap energy; it is not surprising that the per-capita consumption of energy in Australia is 16 times that of Cambodia.

The ready availability of concentrated energy sources from fossil fuels has permitted energy to substitute for labor, and together with technology and access to capital, has driven a more-than-a-century-long upward trend in national productivity, wealth, incomes, and quality of life indicators. A highly productive energy sector with access to high-density energy resources frees the non-energy sectors, allowing them to expand, and fund the industry, healthcare, education, welfare, and myriad of services demanded of modern advanced societies.

But there is a catch; all of our neoclassical economics, our debt-driven growth, modern political economy, and increasingly complex societies have been formulated during a period of rising energy surpluses. What if the *rising* energy surplus experienced over the preceding 200 years is coming to an end?

The enormous energy leverage that used to be available from oil production has been on a declining trend for decades, with much of the "easy oil" already discovered and exploited; capital investment for the global oil industry has tripled in the past 10 years, but oil production has barely increased. Oil producers must drill deeper, operate in harsher environments, and use horizontal drilling and enhanced recovery technologies to produce the same oil that used to be readily available onshore with much simpler operations. Unconventional oil and oil shales now constitutive a rising proportion of production and require much greater investment. China is ramping up expensive coal-to-diesel production, with consequences for greenhouse emissions.

This issue here is not of resources "running out," but a reversal of the trend of increasing energy surpluses. This will inevitably lead to growing energy costs as a proportion of national economies, and the likely consequent slowing down or reversal of improvements in quality-of-life indicators.

Ever since Watt's innovation kick started the industrial revolution there have been four major energy revolutions: the transition from wood to coal and the accompanying steam and steel revolution, the transport revolution driven by oil, the emergence of natural gas as a major energy commodity, and the twentieth century development of nuclear-powered electricity generation. Many people believe we are in the early stages of the fifth revolution: techno-renewables—wind, solar, tidal, and wave.

But therein lies a problem; every prior energy transition led to higher energy density and higher energy utility, but renewables involve a dramatic reversal. The state-of-the-art Spanish Gemasolar solar plant has achieved the holy grail of solar—24 hour electricity generation. With built-in storage and natural gas backup, the plant has a capacity factor of 75 %—on par with base load generators. But the plant requires 300,000 square metres of mirrors with heliostats spread over an area of 195 hectares costing €230 million. To replace just one of Australia's large coal-fired power stations would require building the equivalent of a hundred Gemasolar's on the edge of the desert, along with a 1,000 km of transmission infrastructure. The issue here is not a limitation of technology—Gemasolar represents the pinnacle of 30 years research and development—but what is the economic and energy cost to society?

Unlike the nineteenth century steam revolution, which was able to bootstrap its own energy surplus and expand autonomously, modern techno-renewable energy sources cannot exist in isolation; they are extensions of the fossil-fueled industrial enterprise. To be sure, the strategic use of renewables can "green" the fossil-fueled enterprise slightly, and save the use of valuable fossil fuels for other uses. But when a fuller systems-based approach is taken with more complete boundaries, the resulting energy surplus of solar and other renewables is far lower than frequently asserted, and in some cases may lie below the critical minimum energy-return-on-investment (EROI).

Although renewable energy is often framed around the ideals of sustainability, localization, and simplification, a closer examination reveals that the reverse may be true; the renewables project is a driver of increasing complexity and lowering energy surpluses.

The lessons are clear: the developed world uses too much energy, but the developing world needs far more. A better understanding of biophysical economics, net-energy, and the relationship between energy, ecology, and society is going to be essential if we are going to avoid accelerating the decline in energy surpluses that have been integral to human development over the past 200 years.

Acknowledgment

Many of the ideas for the book were formulated during a vigorous email-based debate following the publication of a paper on solar I published in Sustainability Journal in early 2013, and also shortly after at the fifth Annual Biophysical Meeting in Burlington VT. Hence much of credit goes to those researchers.

Special thanks to Charlie Hall for facilitating much of this, and for the invitation to write the book, and especially for encouraging the development of more rigorous relationships between energy, environment, and society.

Contents

Chapter 1
Introduction: One Million Solar Systems

In March 2013, the number of Australian homes with solar power systems passed the one million mark. For the many who have always advocated for solar, the milestone was a vindication. Indeed, with current solar panel efficiency, and average Australian solar insolation, a square with 31 km sides completely filled with solar panels would collect annual energy equivalent to Australia's current annual electricity requirements. The conclusion could not be more obvious: the future is solar.

The Australian Government's Climate Commission marked the milestone with the release of the report "Australia's future—solar energy", suggesting "the quiet revolution may well mark the beginning of the transformation of Australia's energy systems" [p. 6, (Climate Commission 2013)], and drew a parallel between the transformation enabled by the Internet and the way the solar PV is changing electricity production.

According to the Climate Commission, 2.6 million Australians are "using the sun to power their homes", and solar PV may soon be the most affordable source of electricity for Australians. The Centre for Policy Development's "Going Solar" report, released around the same time, urged policy makers to be embrace solar rather than be guided by "what they see in the rear vision mirror" since "powerful interests will resist such change" (Eadie and Elliott 2013).

But is it really this neat and simple?

Ours is a world run on fossil fuels. The availability of high-density energy sources has provided the energy surpluses to *enable* modernity and the richness and diversity of modern advanced society. To be sure, this has also come at an enormous environmental cost. But we can also fall into the trap of only considering one side of the ledger. What have been the health and respiratory effects of combusting coal on the Australian community? On the other side of the ledger, Australians have a universal public healthcare system with access to pharmaceuticals and expensive surgical procedures—the average Cambodian needs 10 days pay just to afford an asthma inhaler [see Fig. 2 (The International Union Against Tuberculosis and Lung Disease 2011)].

G. Palmer, *Energy in Australia*, Energy Analysis,
DOI: 10.1007/978-3-319-02940-5_1, © Graham Palmer 2014

Solar panels are but one of millions of products that have been enabled by the ready availability of cheap, plentiful energy. But take away the energy surpluses provided by the high-density energy sources, and what is left of the modern industrial enterprise? Is it going to be sufficient to run the mines, transport systems, factories, and myriad of services on solar panels, to make more solar panels?

These issues can be examined with life-cycle analyses (LCA) and net energy analyses. The energy return on investment (EROI), developed by Charles Hall and others in the 1970s (Gupta and Hall 2011), is a ratio of the energy provided by an energy source relative to the energy required in its construction and operation.

At face value, it is self-evident that an energy technology must have an EROI greater than unity or else it will be a net energy consumer. However, Hall et al. (2009) suggests that an EROI of at least 3:1 is the bare minimum when allowance is made of the broader costs of delivering the energy. For example, Hall et al. estimate that an EROI of 10:1 at the oil wellhead translates to an "extended EROI" of 3:1 at the gasoline pump when the energy costs of refining, refinery losses, distribution, and supporting infrastructure are taken into account.

But even the "bare minimum" of three only covers the basic costs of delivering the energy. As noted earlier, the richness and diversity of modern advanced society is only possible because of a much larger EROI, which allows most of the community to be engaged in activities outside of the energy service sectors.

In theory, we could forget about net energy analyses and just look at the monetary costs of energy devices, but the difficulty is that our energy systems are too complex, too subject to subsidies, tax concessions, and uncosted externalities and provide quite different roles. Many people just look at the basic cost of solar and argue that this proves their financial viability, but the basic cost of the panels is really only one component of a large energy system.

This is what this book is mostly about. Rather than doing as most analyses do and note that "intermittency presents challenges to the grid", this book will try to draw out many of the important system-based issues to provide a more informed analysis of solar's strengths and weaknesses.

References

Climate Commission. The critical decade: Australia's future—solar energy. 2013.
Eadie L, Elliott C. Going solar: renewing Australia's electricity options. 2013.
Gupta AK, Hall CAS. A review of the past and current state of EROI data. Sustainability. 2011;3(10):1796–809.
Hall CAS, Balogh S, Murphy DJR. What is the minimum EROI that a sustainable society must have? Energies. 2009;2(1):25–47.
The International Union Against Tuberculosis and Lung Disease. The global asthma report 2011 (2011).

Chapter 2
Quarry Australia: Building Australia on Coal

2.1 Energy Supply 1965–2013: Business as Usual

The remarkable observation of the past 50 years of global energy is the continued dominance of energy by the big three: oil at 33 %, coal at 30 %, and gas at 24 %, which together make up 87 % of global traded primary energy (BP 2012). Australia's proportions of primary energy are 34 % oil, 40 % coal, and 22 % gas (Geoscience Australia 2012). Except for brief pauses during the first and second oil crises and the global economic crisis of 2008, the growth in global energy demand has continued unabated.

Although much of the popular attention of energy policy and policy support over the last two decades has been the role of the techno-renewables—wind, solar, tidal, wave—they still constitute a very small share of global primary energy at around 1 % combined, notwithstanding strong growth from a low base. It is clear that if renewables are to play a meaningful role in a decarbonized energy supply, they will need to replace or displace oil, coal, and gas. The key question is to what degree can renewables take on this role, without further entrenching the role of fossil fuels? (Fig. 2.1).

2.2 Coal: Australia's Saviour or Curse?

Prior to the oil crises of the 1970s, coal's share of global energy was in relative decline, being substituted by gas, and in the case of electricity power generation, nuclear. Writing in 1977, Marchetti (1977) noted that the long-run trend had been a substitution of lower-grade energy for higher-quality energy (i.e. from wood to coal to oil). From the vantage point of the late 1970s and well before global warming emerged, it seemed as though coal's contribution to global energy would fall below 10 % by 2000. The historical lesson here is that left to functional energy markets, the natural trend will be towards higher-density fuels—higher-density

G. Palmer, *Energy in Australia*, Energy Analysis,
DOI: 10.1007/978-3-319-02940-5_2, © Graham Palmer 2014

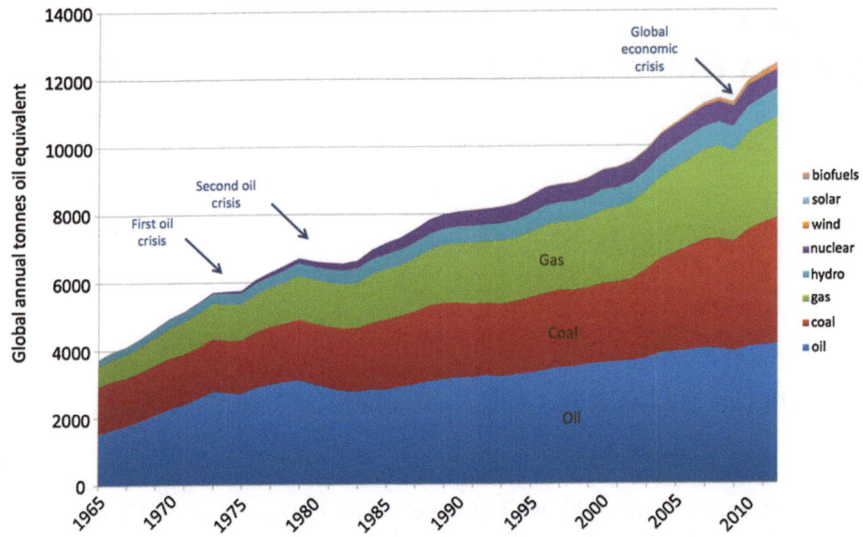

Fig. 2.1 Global traded energy. *Source* BP statistical review of world energy (2012)

fuels result in higher energy content per unit of production, lower costs of transport and logistics, greater utility for end-users, and relatively lower environmental impact.

Vulnerability to oil price shocks, along with the regulatory ratcheting of nuclear, forced the natural trend backwards through the 1980s. Somewhat ironically, the moderation of oil prices by the 1990s meant that coal mining briefly fell out of favour in Australia. Oil companies, including Exxon, sold out of coal mines, leaving the "Big Four" (BHP, Rio Tinto, Xstrata, and Anglo American) to dominate Australia's coal mining industry, making up three-quarters of production (Pearse et al. 2013). In 1960, Australian black coal exports were 1.9 million tonnes, but grew to 96 million tonnes by 1986 due primarily to growth in Japanese demand (Lucarelli 2011).

As of 2013, coal had recovered to 30 % of global primary energy supply, in large part due to the surge in Asian energy demand and steel production. Australian exports of thermal and metallurgical coal combined were 302 million tonne in 2012, with revenue of AUD $48 billion—only iron ore exports are greater at AUD $67 billion (Bureau of Resources and Energy Economics 2013). Overall, Australia is the world's ninth largest energy producer and sixth largest energy exporter on an energy basis, with 63 % of energy production being exported (Bureau of Resources and Energy Economics 2013).

Unlike the USA and Europe, which have suffered from the effects of acid rain and the debilitating effects of localized coal pollution, such as the infamous 1952 London "Great Smog", Australia has largely avoided most of the worst conventional pollution problems normally associated with coal. Although there are adverse health effects such as respiratory and other illnesses due to coal combustion (Castleden et al. 2011), there has been surprisingly little research on the topic in Australia. Perhaps the lack of research interest in the topic is a reflection of the general lack of community concern. For example, the former leader of the Australian Greens and passionate environmental campaigner, Bob Brown, was quoted as saying in 1981 that he would much prefer a coal-fired power station to a hydro dam in Tasmania (Brown 2012). Coal-fired baseload and intermediate-load power generation make up around 75 % of national electricity production.

However, more recent serious community concerns relate to coal mining for export. The recent large-scale expansion of coal mining for export markets is now impinging upon rural communities and prime agricultural lands in New South Wales (NSW) and Queensland. Not surprisingly, the defence of local communities is difficult against the relentless pressure applied by large miners (Pearse et al. 2013)—exports are projected to grow to 581 million tonne by 2035. Australian exports account for around 27 % of world coal trade, most being sourced from NSW and Queensland. Japan is the largest importer of Australian coal, followed by China, Korea, and India. Victoria's large reserves of lignite are significant, but the high moisture content renders the product unsuitable for export (Fig. 2.2).

The growth in export markets of coal and iron ore has been drivers of a strengthening Australian dollar, placing greater pressure on other exports and exposed industries, including agriculture, manufacturing, tourism, and education services. This leads to the "Dutch disease"—a higher Australian dollar reduces the relative export competitiveness and increases demand for labour and services, driving up costs. However, unlike earlier mining booms, the floating exchange rate (which was introduced in 1983) and more flexible goods and labour market provide a means to better accommodate booms (Battellino 2010). For example, a higher dollar also reduces the cost of imported oil for Australians.

The first serious consideration of supplanting coal-fired electricity generation was in the late 1960s when serious consideration was given to nuclear. A large power reactor was planned for Jervis Bay, NSW, in the late 1960s, and construction preparations had commenced—the concrete footings are now being used as a car park for the local Jervis Bay surfing community (Owen 2011). The project was eventually cancelled after a change of Prime Minster and a comprehensive cost review by the Treasury, which revealed the relative cost advantages of coal-fired generation against a nuclear plant (Cawte 1992).

Around the same time and in response to strong demand growth for electricity, the chairman of the State Electricity Commission of Victoria (SECV), W. H. Connelly, envisaged the gradual expansion of Victorian nuclear capacity based on 500–600-MW class units (Edwards 1969). In 1970, Sir Philip Baxter was more optimistic, suggesting that twenty 500-MW units might be started by the late 1980s Australia wide (Baxter 1980 and Baxter and Griffiths 1969).

From the 1970s through the 1980s, the SECV initiated similar cost reviews and came to the same conclusion; coal presented a better economic proposition. The fact that coal was still chosen over nuclear during the emerging nuclear age is significant and testifies to coal's low cost, traditional social licence, and the substantial investment and expertise gained in exploiting

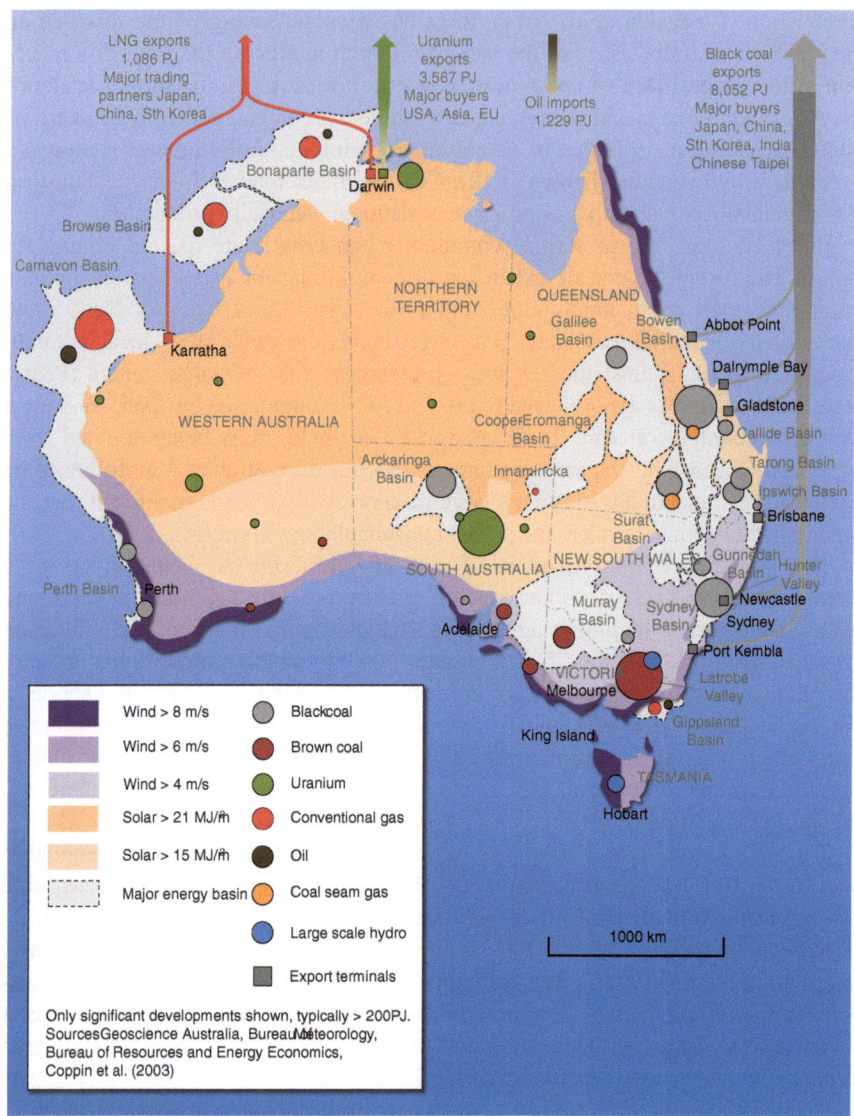

Fig. 2.2 Australia's energy resources. *Source* Geoscience Australia (2012)

Victoria's brown coal (lignite), notwithstanding more recent changes in community sentiment in relation to concerns over the high emission intensity of coal-fired electricity.

Australia has a unique role as being a major uranium supplier without participating in other elements of the nuclear cycle—Australia's known uranium resources constitute 31 % of the global total. Uranium oxide (U_3O_8) comprises a modest share of Australia's export volume and revenue, at 7,499 t and $670 M, respectively. Although the tonnage is modest, it represents 20 % of Australia's energy exports in energy terms. Notwithstanding a reassessment of nuclear post-Fukushima, there are 10 countries in Asia planning an expansion of nuclear capacity, with China also planning a long-term shift to fast-neutron reactors. In the main, this reflects a commitment to energy security with rapidly expanding demand.

Even in the event of the removal of current Australian regulatory impediments to nuclear power, the implementation would pose a number of challenges (Owen 2011), especially since the "front-loaded" cost structure is more suited to state-owned utilities. Nuclear power provides the cheapest electricity in most OECD countries, but private investors in liberalized markets require lower risk and quicker returns—indeed, the expansion of low-capital-cost gas turbines is a reflection of the current market conditions.

Australia is also gas rich, with liquefied natural gas (LNG) exports of AUD $12 billion, based on large reserves in the Carnavon, Browse, and Bonaparte Basins off the north-west shelf and domestic supplies sourced from the more mature Gippsland basin.

It should be borne in mind that Australia, and Victoria in particular, has invested enormously in coal-fired generation. Victoria's brown coal resources were untapped for the first century of the pioneer settlement's development, and it was not until the 1920s that the SECV, led by Sir John Monash using German technology, was able to capitalize on the resource. Victoria had relied on black coal imported from NSW until the middle of the twentieth century. NSW possessed large deposits of high-grade black coal, which had been used since the mid-1800s. This gave NSW a considerable economic advantage in manufacturing, as well as iron and steelworks. Victoria's Tertiary-era lignite was always considered a second-rate resource, primarily because of its high moisture content, but was developed in response to unreliable supply from NSW due to ongoing industrial disputation.

Even today, a visit to Victoria's Latrobe Valley reveals a long-term tradition in Victoria's low-grade brown coal and a constituency ready to defend it. The high emission intensity of lignite is, theoretically at least, to be addressed with carbon capture and sequestration (CCS). Pilot plants have been successfully trialled, and the Latrobe Valley's proximity to the Gippsland basin provides an ideal site for sequestration. The remaining piece of the puzzle is working out how to fund the multi-billion dollar commercial-scale development, along with the higher operating costs in a competitive electricity market—the Federal Government's Clean Energy Finance Corporation excludes CCS from funding.

The most advanced of the large-scale Australian CCS projects was to be the "ZeroGen" integrated gasification combined cycle (IGCC) and (CCS) power plant and storage facility in central Queensland. The capacity was to be 400 MW at a cost of AUD $4.3 billion. The project was cancelled in 2010, coincidentally around the same time as a major expansion of Queensland coal exports.

In the absence of a credible carbon-pricing scheme, it is difficult to see the viability of CCS—indeed, the most effective strategic position for the coal industry is to support pilot-scale projects without committing to large-scale commercialization. Unless there are off-the-shelf solutions available, governments are hamstrung in legislating for carbon capture (Palmer 2009), and in any event, even with a modelled carbon price of AUD 2010 $55 per tonne CO_2-e, some of the highest emission coal-fired generators in Australia, including Hazelwood, Loy Yang, and Yallourn, have amongst the lowest short-run marginal cost (ACIL Tasman 2009).

As a country that is coal rich, but with declining oil supply, there have been proposals to develop coal-to-liquids (CTL), which will result in a high-cetane, low-sulphur diesel, but with high life-cycle greenhouse emissions. Just as the ramp-up of fossil fuels from the nineteenth century followed a trend of increasing energy density and quality, it seems plausible that the likely ramp-down during the twenty-first century may follow a symmetric trend, with coal taking on a greater burden, including liquids fuels.

This raises the broader issue of the impact the declining energy surpluses will have on society. Using data on the US gas and oil industry, King and Hall (2011) have shown that there is a long-run correlation between EROI and energy prices, such that a declining EROI (and therefore reducing energy surpluses) is correlated with rising energy costs. This emphasizes a physical basis to energy costs alongside market-based costs and raises the question as to whether the long-run risks of rising energy costs are adequately reflected in short-run market-based prices.

King (2013) draws attention to the relationship between US petroleum expenditure as a proportion of GDP and recessions, observing that a rise above 4 % is correlated with recessions (shown as dark areas—see Fig. 2.3). But Kilian (2009) also raises the issue as to how much monetary policy, alongside the oil price shocks during the 1970s, contributed to stagflation, particularly the role of rising interest rates to curb inflation. The shift from Keynesianism to Friedman's monetary policy—central bank targeting of inflation with interest rates—confounds systematic analyses. Kilian notes that rising oil prices may coexist with a strong economy provided there is strong global aggregate demand. It seems likely that EROI is an underlying factor, but establishing a robust relationship between EROI and economy-wide effects remains allusive.

The CTL process was invented in coal-rich Germany in the 1920s and is named after the German researchers Franz Fischer and Hans Tropsch from the Kaiser Wilhelm Institute. In response to supply embargos during World War 2, German production reached 124,000 barrels per day in 1944. Similarly, South Africa undertook CTL in response to apartheid-era sanctions. Sasol currently produces about 150,000 barrels per day (Gibson 2007).

Both examples of large-scale CTL point to the fact that CTL is an expensive undertaking, but the process produces an easily transported and converted fungible product that can be blended with conventional diesel and integrates into existing infrastructure.

Archibald (2008) suggests the costs are comparable with deepwater oil and LNG projects. A proposal by Monash Energy using Victoria's lignite was costed at AUD $5 billion to produce 60,000 barrels of liquid fuels per day, equivalent to

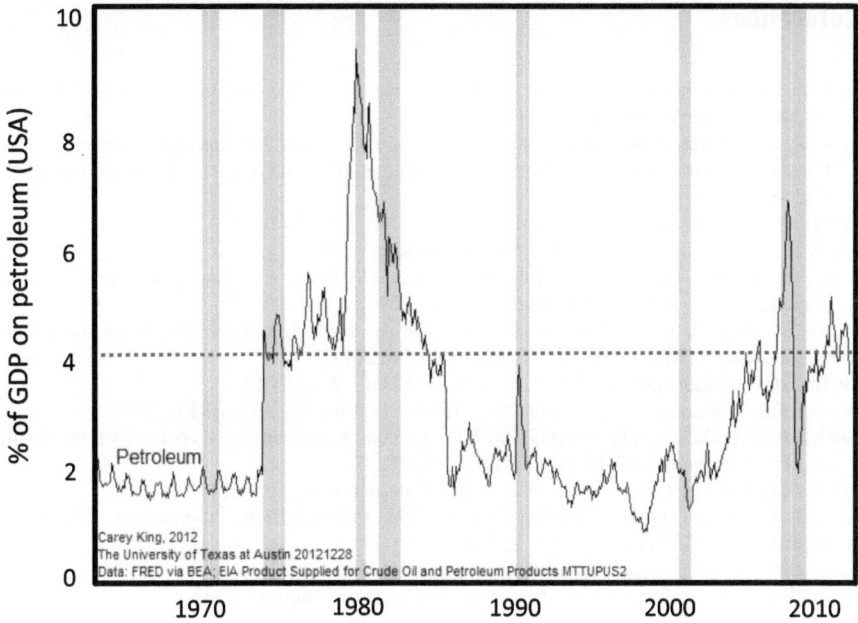

Fig. 2.3 Correlation between petroleum costs as a proportion of GDP and recessions (*dark areas*) in the USA. A rise above 4 % appears significant. *Source* King (2013)

about 6 % of Australia's current demand. It included provision to sequester the production-related CO_2 emissions in the Gippsland basin (Australian Government 2007).

ABARE suggests a capital cost of US $50–70,000 per barrel of daily capacity, which translates to a cost of AUD $3–5 billion based on a capacity of 60,000 barrels per day. Based on these figures, ABARE suggests a long-term oil price above US $40–45 per barrel is required. Although technical risks are small since the suite of technologies are proven, the scale of the investments, which can potentially "break the company", it is not surprising that CTL has not emerged commercially in Australia. Risks include legislative risk associated with environmental conditions, carbon pricing, oil price volatility, and exchange rate uncertainty.

In contrast, state-run coal companies are developing six CTL plants in China, including one direct-coal-to-liquid (DCL) project (Wu 2012). The strategy of CTL is a reflection of the importance of energy security given that China imports around half its oil and is natural gas poor.

Australia has also developed coal seam gas (CSG), most of which has come from Queensland, but production is also growing in NSW. CSG expanded from 2 % of total gas production in 2002 to 11 % in 2010 and is continuing to expand. Like the USA, Australia also possesses very large reserves of tight, or shale gas, but exploration is still in its infancy. Known resources are in low-permeability sandstone reservoirs in the Perth, Cooper, and Gippsland basins.

References

Archibald D. Submission to the senate committee—the oil supply—demand outlook and the case for an Australian coal to liquids industry. 2008.

Australian Government. Inquiry into Australia's future oil supply and alternative transport fuels: final report. Australian senate standing committee on rural and regional affairs and transport, Canberra 2007.

Battellino R. Mining booms and the Australian economy. RBA Bull. 2010; 63–69.

Baxter P. Atomic power by the 1980s. The Age March 19 1970.

Baxter J, Griffiths D. Nuclear power. In: Raggatt H, editor. Fuel and power in Australia. Melbourne: F. W. Chesire Publishing; 1969. p. 108.

Brown B. ABC—Q & A—Bob brown joins Q & A. Available online: http://www.abc.net.au/tv/qanda/txt/s3478779.htm (2012). Accessed 1 Jan 2013.

BP, British Petroleum. Statistical Review of World Energy 2012, 2012.

Bureau of Resources and Energy Economics. Energy in Australia 2013, BREE. 2013.

Castleden WM, Shearman D, Crisp G, Finch P. The mining and burning of coal: effects on health and the environment. Med J Aust. 2011;195(6):333–5.

Cawte A. Atomic Australia. Sydney: New South Wales University Press; 1992.

Edwards C. Brown power—a jubilee history of the state electricity commission of Victoria. Adelaide: The Griffin Press; 1969. p. 235.

Geoscience Australia. Australian energy resource assessment. 2012.

Gibson P. Coal to liquids at Sasol—Kentucky energy security summit. 2007.

Kilian, Lutz. Oil price shocks, monetary policy and stagflation. Centre for Economic Policy Research, 2009.

King CW. Net energy principles for understanding if energy production is an economic constraint. Webber energy group symposium 2013.

King CW, Hall CAS. Relating financial and energy return on investment. Sustainability 2011; 3(10):1810–1832

Lucarelli B. Australia's black coal industry: past achievements and future challenges. Program on Energy and Sustainable Development. 2011.

Marchetti C. Primary energy substitution models: on the interaction between energy and society. Technol Forecast Soc Chang. 1977;10(4):345–56.

Owen AD. The economic viability of nuclear power in a fossil-fuel-rich country: Australia. Energ Policy. 2011;39(3):1305–11.

Palmer G. Out of sight or out of time: the future of carbon capture. Dissent 2009;43–48 (Spring 2009).

Pearse G, McKnight D, Burton B. Big coal Australia's dirtiest habit. Montgomery: NewSouth Publishing; 2013.

Tasman ACIL. Fuel resource, new entry and generation costs in the NEM, 0419-0035. Melbourne: AEMO; 2009.

Wu N. Coal to liquids: why and how it makes the case in China. Available at SSRN 2190078. 2012.

Chapter 3
Towards Optimized Complexity: Integrating Intermittency

3.1 Josep'h Tainter's Complexity Spiral

Much of the discussion of renewable energy is posited on the ideals of the "soft energy path", sustainability, localization, and simplification. Yet when these ideals are unpacked, it becomes clear that realizing them will require a radical increase in complexity.

According to Tainter's thesis (Tainter 1990), societies become more complex as they try to solve problems, including new layers of bureaucracy, infrastructure, and social class. In the initial stages, the complexity helps resolve the problem, but additional complexity is subject to the law of diminishing returns.

For example, specialization improves labour productivity and has been a primary driver of income and economic growth. Examples include the trend from general medical practice to surgeons who specialize in only a handful of surgical procedures, or engineers who practice within a narrow field within a particular industry. Tainter notes that whereas hunter-gatherer societies contain not more than a few dozen distinct social personalities, modern advanced societies recognize 10,000–20,000 unique occupations.

But specialization also has a cost; investment in education is growing, more students undertake advanced education, the productive phase of graduates is delayed and post-education specialization carries further costs. Although advanced education and specialization may continue to provide a net benefit to society, it is subject to a trend of declining marginal productivity.

Climate change, depletion of non-renewable resources, and a decline in EROI are also examples of problems that are subject to increasing complexity. Yet the often-assumed solution of the "soft energy path" is itself part of the problem of increasing complexity and a declining marginal productivity. An examination of the various "100 % renewable" plans provides an opportunity to assess the increasing costs of increasing complexity.

3.2 100 % Renewable Energy Plans

It has become common to discuss the evolution of low-emission electricity systems as a suite of geographically and technically diverse energy sources. Combined with plausible energy efficiency goals, demand management, and emerging storage options, it is should be possible to wholly, or nearly wholly, run an electricity grid on renewable energy sources. Indeed, hypothetical plans describing such systems have become a virtual industry in recent years. These have been variously undertaken by environmental NGOs, academics, and government renewable energy agencies.

Recent Australian examples include Australian Beyond Zero Emissions (Wright and Hearps 2010), Elliston et al. (2012), and Australian Energy Market Operator (AEMO) (2013). There have been many other plans undertaken internationally, for example World Wide Fund for Nature (WWF 2011), Zero Carbon Britain (Kemp 2010), Greenpeace International and European Renewable Energy Council (Teske 2010), Jacobson and Delucci (Delucchi and Jacobson 2011) and Budischak et al. (2012).

Although every report is a unique undertaking, they mostly share the same common themes; concentrating mostly on a specific electrical grid or region, a simulation is run with a combination of wind, solar, hydro, biofuels, and storage to demonstrate that renewables can power society.

3.3 Intermittency Managed Through Complexity and Exploiting Synergies

The underlying theme of 100 % renewable plans is the assumption that through increased complexity, an optimal set of synergies can be discovered and exploited. An example might be the hypothesized synergy between solar PV and electric vehicles (EV), in which the grid benefits from having storage embedded within the distribution network, while EVs benefit from accessing low-cost embedded solar power. The optimum charge regime for a fleet of EVs and PV systems would be calculated in real time by a smart grid.

Another might be combining geographically diverse solar farms with wind farms, such that it will usually be windy or sunny somewhere. These plans are usually undertaken as desktop simulations, and although different groups use their own methodology, most plans use a variant of the general simplified process as follows:

1. Assume various solar plants, wind farms, and other renewable energy infrastructure will be located in specific geographic locations
2. Using detailed hourly or half-hourly weather and climate data for the plants, run an hourly simulation covering a full year and compare to historical annual demand for the same year

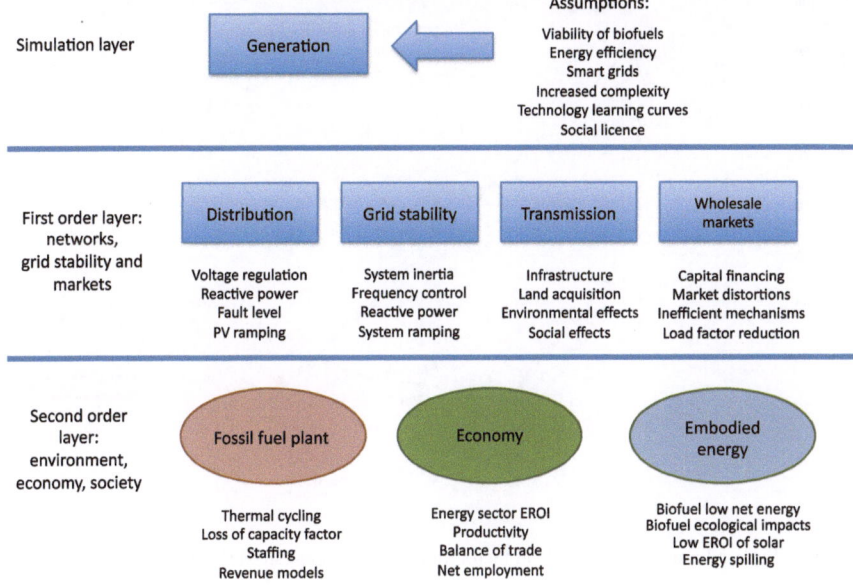

Fig. 3.1 The hierarchy of renewable energy plans. Most plans only consider the *top* layer

3. Fill in shortfall in demand with a combination of hydro, storage, biofuels, or backup fossil fuels
4. Run simulation again, and reiterate until a satisfactory solution is found that satisfies historical demand for every hour for the whole year for a given reliability standard

The difficulty is that the plans operate solely within the shallow "simulation layer" shown in Fig. 3.1 and require many theoretical critical assumptions. With few exceptions, little consideration is given to the deeper first- and second-order layer issues. This chapter will consider some of these issues and try to draw out the additional challenges.

3.4 Reduced Load Factor of Electricity Systems with PV

Figure 3.2 is a stylized graph of system load factor or generator capacity factor for various electricity grids and generator types. In large electricity grids, most of the system energy has traditionally been supplied by large, high-utilization baseload units, which provide the lowest average cost generation when operated with a high capacity factor. In Australia, this has traditionally been supplied by coal-fired generation using locally mined coal (the issue of greenhouse abatement is discussed

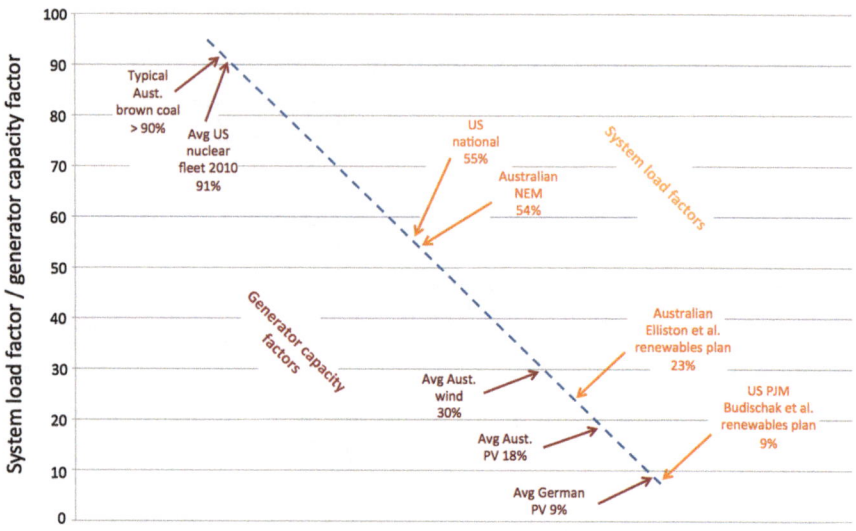

Fig. 3.2 Stylized graph of system load factor and generator capacity factors

further in Chap. 6). The annual system load factor of the entire Australian National Electricity Market (NEM) is also shown. A high system load factor implies the efficient use of plant.

The increasing use of intermittent generation represents a shift away from lowest-cost electricity delivered by large, high-utilization units towards lower-utilization electricity generation. Indeed, there is broad consensus that a shift to intermittent sources of electricity will require greater use of quicker response dispatchable generation, such as gas turbines as well as large-scale storage. A shift to lower capacity factor generation will necessarily lower the system load factor and implies that a significant proportion of generation units are idle for most of the time with obvious consequences for costs.

Much of the decline in system load factor is a consequence of the fact that the availability factor for renewables is a function of both technical reliability and temporal availability of the natural resource, whereas conventional generation reliability is solely a function of technical reliability of the machines.

Perhaps one of the starkest examples of the poor load factor of high-penetration renewables is the Budischak et al. (2012) renewables simulation of the PJM regional grid in the USA. Applying a small discount to allow for their optimistic capacity factors and adding a conservative reserve margin, the resulting load factor is calculated at around 9 % based on the "99.9 %" scenario from Budischak et al. Table 3. Additionally, 51,900 MW of storage capacity is required. Although the study is based on the "least-cost" simulation, it is apparent that a system with a net idle factor of greater than 90 % represents an extraordinarily wasteful use of capital.

3.5 The Illusion of Synergies

The reduction in load factor of high-penetration renewable schemes is a consequence of the illusion of synergies. The issue here is one of the probabilities. Diversifying both the geographical span and type of renewable generators increases the probability that there will be sufficient power being generated at any time, but the availability and capacity factor of intermittent generation is extremely low relative to system reliability standards.

Although there are often instances of low-level correlation between various types of renewables (e.g. greater average wind velocity during winter or at night when solar is at a minimum), the correlations are very weak relative to the defined reliability standards imposed on system operators (the Australian standard is essentially 99.998 % reliability). In particular, winter performance is most problematic, since even Australian regions with excellent annual solar radiation can experience extended periods during winter when little or no solar is available (Trainer 2010b), a problem exacerbated with concentrated solar thermal, which has a thermal threshold before electricity is generated.

Hence, high-penetration renewable plans inevitably include overbuilt capacity to cope with the weak correlations and must rely on backup and storage to "fill the gaps". This leads to the problem of overbuild with an unaffordably high investment/GDP ratio (Trainer 2012), and therefore a diversion of resources from the non-energy sector.

Another example is the hypothesized synergy between solar PV and electric vehicles (EV). Although at face value the synergy can seem to work at a local scale, on a system-wide scale it is not obvious that the synergy is as clear as first seems (Palmer 2013 and Trainer 2012, 2013).

Instead, a more obvious synergy is between baseload, which can provide low-cost and predictable year-round off-peak power for charging when most vehicles are parked at home, while underpinning the load factor for baseload generators in the event of the electrification of the Australian motor car fleet. Although not as highly correlated, the use of EVs to absorb excess wind at night-time may improve the value of wind energy (Lund and Kempton 2008).

The problem with solar PV-based vehicle charging is that available solar supply will be inversely related to the preferred charging regime (i.e. there will be no PV power available for night-time charging when the grid has spare capacity, but during the 3 or 4 h centred on solar noon, few motorists will want to "fill their tank" at the peak daytime tariff).

Further, although the significant distributed storage made available by a fleet of parked and plugged-in EVs may offer a valuable niche role for network support during critical peak demand events, it would make little sense to cycle a significant proportion of system energy through EV batteries to support intermittent generation; EV batteries are cycle limited, and the resulting degradation will reduce the longevity of the expensive EV batteries and prove uneconomic for motorists.

Regardless, when a fuller account of the lifecycle of EVs is included, the high embodied energy content of batteries results in only marginal emission gains over an equivalent-sized fuel-efficient conventional vehicle (Crist 2012; PE International 2012). In the case of biofuels, sugar cane-based ethanol appears more promising than corn based, but to date, the only moderately "sustainable" example of biofuels has been the example of Brazil (Cleveland and Saundry 2007). Ethanol may play a greater niche role in transport given the right conditions. Once again, the increased diversity of energy systems may provide some benefits, but it also increases the complexity of the energy system, and therefore, the energy overhead of maintaining increased complexity.

3.6 Redefining Baseload

There have traditionally been two overlapping ways to define baseload power. The first considers the *demand* profile and is defined as the minimum power demand on the electricity network, usually the minimum overnight load. The second considers baseload in terms of *supply* and is generally understood to mean low marginal cost generators that run continuously at utilization rates of greater than 70 % such as coal or nuclear (Institute for Energy Research 2012).

In conventional grids, the intended role of baseload generation is to provide the bulk of the energy requirements at lowest cost, with higher marginal cost peak and intermediate load providing load following (Anderson 2012). In smaller, isolated grids, baseload is not essential, and diesel generation can take on the primary role.

Since all of the renewable plans rely heavily on intermittent generation, the role and definition of baseload have become blurred in recent years. Typical of these is Diesendorf's "baseload fallacy" (Diesendorf 2007b), which follows a process of deductive reasoning to conclude that a combination of efficiency, renewable energy, and storage can substitute for baseload. In more recent studies, Diesendorf (2007a) and Elliston et al. (2012) argue that baseload can be thought of as a *system*, rather than an individual generator unit, suggesting that we require a "radical twenty first century reconception of an electricity supply-demand system".

On the other hand, Beyond Zero Emissions prefer to maintain the conventional use of the term but refer to "baseload solar" in relation to concentrated solar thermal with storage. Sovacool (2009) argues that the concept of baseload is as much about the "social, political, and practical inertia of the traditional electricity generation system", citing the large penetration of wind in Denmark as evidence of the potential for intermittent generation to displace baseload. Similarly, Archer and Jacobson (2007) suggest that a significant proportion of the capacity of interconnected wind farms can be used as baseload power.

Much of the discussion of baseload is really referring to what is more accurately depicted as dispatchable power (i.e. machines that are controllable with a high availability factor). Although baseload is not *ipso facto* an essential element of electricity generation, it has a number of system and stability roles that are unique to large, synchronous generators, discussed later.

3.7 The Synchronous Australian National Electricity Market

The Australian NEM spans five states and 4,500 km and has around 260 registered generators (Australian Energy Market Operator (AEMO) 2010), all of which when online are synchronous machines spinning in near-exact synchronization at close to 50 Hz across the network [the island state of Tasmania is connected via the non-synchronous Basslink DC link (Electricity Supply Industry Expert Panel 2011)]. A synchronized grid implies that the rotors of all generators are spinning in exact synchronization throughout the entire network. In Australia, the grid frequency of 50 Hz corresponds to a rotor speed of 3,000 RPM for 2-pole generators, or for example 500 RPM for large 12-pole hydro generators. Indeed, nearly all global electricity is produced by synchronous generators driven by rotary turbines; in 2010, 96 % of global electricity was generated by thermal or hydro plant (International Energy Agency (IEA) 2012), with nearly all being either steam, gas, or hydro turbines (Smil 2005).

The synchronization of the network can be considered by imagining a road train of 100 vehicles of various sizes, which are tethered together in a long line, driven along an undulating highway at constant speed. In much the same way, generators become "locked" by the interaction between the generator magnetization and grid cycle when they are connected online and synchronized.

But maintaining the correct speed can be challenging and requires sophisticated monitoring and feedback mechanisms to ensure stability of the entire 4,500 km-long "machine". Indeed, in the early period of electricity generation, generators operated in electrically isolated networks since the parallel operation of generators used to be a daunting engineering problem. But from the 1930s, the theory of parallel operation of generators in large networks was established (United Nations 2006).

In the road-train analogy, every vehicle should be applying power, but if each car used its own cruise control to try to meet the target speed, some vehicles would be working against each other; it would be counterproductive to have some vehicles applying full power while others coast or brake.

The method used to measure the supply–demand balance in synchronous electricity networks is through monitoring the grid frequency. Every electricity outlet throughout the NEM has exactly the same frequency at every instant, so it is possible to measure in real time whether the grid is slightly overpowered or slightly underpowered.

When the aggregate load increases, all the generators slow down due the additional load, so in response, a little more power needs to be applied to the generator rotors through the turbines. Then, conversely as the load declines, the power needs be throttled back.

The way in which each generator calculates the required power to apply is through a control process known as "droop", which is a feedback process that correlates the frequency to the required power. For example when the target frequency is 50 Hz, this might correspond to a generator power of 90 %, but if the frequency speeds up to 51 Hz, the power may need to throttle back to 85 %.

Conversely, the power may ramp up to 95 % when the frequency lowers to 49 Hz. This process occurs in real time, in conjunction with the wholesale electricity market, and is a highly dynamic process. In the road-train analogy, the baseload generators are equivalent to diesel trucks that maintain the road train and do most of the heavy lifting, while the smaller gasoline vehicles power up or power down to keep the road train at a steady pace.

3.8 System Inertia

An equally important parameter in electricity networks is the system inertia. When transients occur in the supply–load balance, some of the mechanical rotary inertia in the large turbines and generator rotors will be expended to dampen the instantaneous frequency change, giving generators time to respond to the changing balance without loss of stability. The damping provided by the inertia underpins the stability of the network and allows generators to gradually respond through droop. In the road train example, the inertia is simply the total mass of all of the vehicles; in the case of electricity networks, the inertia is a function of the momentum of the rotor and turbines sets, along with the stored energy in the inductive and capacitive elements.

3.9 Non-Synchronous Generation

Solar PV and wind generation are non-synchronous generators, meaning that they inject power into the grid through electronic inverters rather than through a spinning rotor (in contrast, concentrated solar thermal injects power through a synchronous turbine/generator set). As such, they do not add inertia or transient damping to the system—indeed, the rapid ramping of PV due to cloud flicker, the rapid onset of cloud, or the arrival of sunset worsens rather than improving transient behaviour.

Wind exhibits much less ramping on a shorter time scale (seconds to minutes), but collectively across the Australian NEM induces a much higher ramp rate on an hourly time scale than normal demand-driven ramping, relative to capacity.

Whereas baseload generators are the equivalent of an ocean liner pushing through ocean swell and capable of absorbing transients, the nation's collective PV and wind generation is equivalent to a surfboard sitting on top of the ocean chop; they can only follow the grid frequency undulations, not dampen them.

At a low penetration of low-inertia plant in a large grid, this is of little consequence since the NEM has more than adequate inertia to cope with low-penetration non-synchronous plant. However, in Tasmania, the increasing penetration of wind in an otherwise small grid has forced the grid operator to formulate strategies to ensure network stability with the relative loss of inertia. For example, the Tasmanian Energy Department (Tasmania Department of Infrastructure Energy and Resources 2010) has run hydro generators in synchronous condenser mode, noting the additional costs due to increased wear and cavitation and more frequent starting and stopping.

It is possible to install dedicated synchronous condensers (essentially generators without a prime mover) to provide the inertia for non-synchronous generators, or use battery storage with electronic controls to create "virtual inertia" to emulate physical inertia. However, all of these remedies incur a cost, and since the market has not traditionally required dedicated inertial support, there is no market signal under the current regulatory arrangements.

3.10 Offloading or Heat Rate Losses of Conventional Thermal Generation

At a low penetration of intermittent generation, the role of conventional generation is largely unchanged since intermittency is accommodated in the same way as demand variability through the normal operation of the wholesale market and flows across state inter-connectors. As the penetration of intermittency increases, baseload generation may be required to take on an increased load balancing role. This in turn may force operation away from their optimum heat rate or result in offloading.

In grids, such as the US ERCOT system in Texas, studies have found that some of the emission gains of wind may be dissipated through heat rate losses of the coal-fired generation (Kaffine et al. 2011). However, the actual losses are specific to the particular grid. Where these issues have been examined in Australia, modelling suggests these effects will be minor at a moderate penetration of intermittent power since the grid is already configured for variability. This applies particularly to solar PV, which only generates during the day and displaces intermediate and peaking plant. At increasing penetrations of intermittent generation, particularly from around 5 to 10 % of annual energy, these effects may become more pronounced and result in a material reduction in expected abatement (Inhaber 2011).

3.11 Retaining Fossil Fuel Plant for Reliable Capacity

As part of a shift to a high-penetration renewables scenario, baseload generators may be taken offline for much of the year, but retained to provide the needed capacity during the high season, which in Australia is the summer period. SKM MMA (2011) conducted an analysis into the viability of taking two of Victoria's high-emission intensity coal-fired generators offline for most of the year, while maintaining them in preservation mode for restart during summer to meet high demand. This was conducted as part of the Garnaut Review (Garnaut 2008) into introducing carbon pricing.

Preservation mode requires maintaining corrosion protection for the steam circuits, boiler components, and turbines and protecting the electrical generator against moisture ingress. Much of the machinery and equipment would need to be operated periodically in order to maintain lubrication, limit moisture ingress, minimize corrosion, and ensure operational availability. An increase in metal fatigue would likely occur due to increased shut-ups/shutdowns, increased thermal cycling, and greater susceptibility to corrosion.

The most serious challenge was deemed to be staffing arrangements, with issues of retaining experienced staff, training rehired staff, and managing work agreements. This would likely have implications for staff morale, retention salaries, retraining costs, and workplace contractual obligations. The modelling projected a decline in capacity factor from the usual baseload factor of 80–90 % down to 3–8 %, although the generators would be expected to be online for a greater time.

The financial model would likely require a complete revision, since the capital and fixed operating costs could not be recovered from the routine wholesale energy market operations. Further, the shutdown would preclude the plants from participating in the provision of ancillary services for most of the year. Hence, costs would need to be apportioned to generators sitting idle for much of the year.

Similarly, the Budischak et al. (2012) renewables study of the PJM regional grid in the USA, assumed that 28,300 MW of fossil fuel plant would be retained in their "99.9 % scenario", equal to just under the average grid demand, but accounting for a miniscule 0.02 % of system energy and hence operated at well under 1 % capacity factor.

Although the study acknowledged that such an arrangement may be uneconomic, it does not provide a satisfactory response to managing the dilemma of requiring capacity to reliably meet demand, but removing the conventional revenue stream required to allow the generator to remain profitable. These examples demonstrate the gulf between desktop simulations and the complexities of managing real-world scenarios.

3.12 Altered Voltage Profile Due to Distributed Solar PV

Although it is often assumed that solar PV reduces the costs associated with distribution (Eadie and Elliott 2013), the altered voltage profile and variability of solar PV represents an additional cost to the distribution network. In some cases, depending on the relationship between solar and peak demand, there may be a small to moderate reduction in peak demand as a consequence of PV, potentially leading to network augmentation deferral (see Chap. 4). In most regions, however, the peak load on the highest demand days occurs during or after sunset, mostly in response to air conditioner loads after people arrive home from work. Hence, PV does not provide a meaningful reduction in annual peak load without the addition of integrated storage and therefore provides little or no reduction in distribution costs (Palmer 2013).

The additional costs associated with distributed PV generation are due to the change in the voltage gradient along feeders and are not applicable at a low local penetration of solar. The problem is that the system is designed around unidirectional power flows and is intended to accommodate the predicable voltage gradient that occurs on residential feeders from substations; the nominal voltage in Australia had been 240 V from 1926 but was changed to 230 V in 2000 to align with international standards (Halliday and Urquart 2011). Normal variations in load along the low-voltage feeder cause voltage variations within prescribed limits; voltages will typically be higher at night-time and lower during periods of high demand. Voltage regulation is achieved locally through transformer on-load tap changers (OLTC) at electrical substations. The injection of power along the feeder raises the voltage gradient and can contribute to three-phase imbalance and is particularly problematic when it occurs at the tail end of the feeder during periods of high solar insolation and low demand, such as during late mornings.

Feeders are described as "weak" if there is high impedance on the line caused by low loads, leading to higher voltage variance in response to a change in load (Bindner 1999). Australia has weaker grids and a different network structure to the USA and Europe, and as such, network limits to solar PV may be reached sooner than in comparable countries (Sayeef et al. 2012).

Distributors are required to maintain minimum standards of performance and reliability and will reject solar applications in instances where the injection of power may reduce reliability of supply; indeed, this has already been occurring in Australia (Sayeef et al. 2012).

There is no fundamental technical restraint on PV injections; the problem is one of cost. Technical solutions include the replacement of OLTC transformers with automatic tap-changing transformers, the greater use of three-phase inverters, smart grid integration, bidirectional (four-quadrant) inverters, or the future development of new distribution architectures (e.g. "Active Networks", "MicroGrids", and "Virtual Power Plants").

It is difficult to estimate the marginal cost of accommodating the altered voltage profile since significant costs are only likely to be incurred beyond a few percentage of local PV penetration, but the pro-rata cost of transformer upgrades or conversion to three phase could substantially add to the basic cost of a household PV system with significant consequences for the net-financial viability and EROI (Palmer 2013). The potential scale of upgrades is significant considering that there are around 550,000 distribution transformers in Australia with a total power handling capacity of 100 GVA (Blackburn 2007).

3.13 Smart Grids and Demand Management

The development of a "smart grid" is generally considered an essential component of high-penetration renewables (Bower et al. 2012) and is an evolutionary process that will evolve over many decades (International Energy Agency (IEA) 2011). Victoria is the first Australian state to implement a "smart meter" program with time-of-use (TOU) pricing, which should eventually lead to a moderation of a declining system load factor (Simshauser and Downer 2012). Other benefits include a reduction in meter reading costs and improved fault detection.

There are potentially significant benefits in improving the information management of the electricity network, including flattening the demand curve and enhanced consumer and network operator feedback (Australian Productivity Commission 2013). Demand management for a few critical peak demand events in the absence of smart metering has already been widely used by large energy consumers in Australia for decades. Perhaps the simplest demand management tool, off-peak switching for domestic electric hot water services, has been available in Victoria since the 1930s. The issues here are cost, equity, and capturing the potential benefits; the early projected cost of the Victorian roll-out of $800 million eventually rose to $2.3 billion.

For example, high air conditioner use on the hottest days is one of the key drivers of a reduced system load factor and rising network augmentation costs. In theory, time-of-use pricing during critical peak demand events is meant to discourage the use of air conditioners and other minor loads during these periods, and trials have confirmed the potential for demand reduction in response to consumer and price feedback. But it is not surprising that these events occur on the hottest days and that most householders will elect to use their air conditioner on those days of over 40 °C (104 °F). Indeed, even frugal consumers who generally use their air conditioning sparingly will want to use the cooling on the hottest days; this is the whole point of buying an air conditioner.

Given high enough TOU tariffs, consumers on fixed incomes such as pensioners and other low-income earners will choose to turn their cooler off, and conversely, high-income earners will simply bear the cost. This of course introduces issues of equity and health, particularly where it relates to the elderly and vulnerable, and also raises the issue of the degree to which ordinary consumers are expected to understand complex tariff arrangements.

A cost recovery model based on the greater use of a fixed demand-based charge would also align actual costs with consumer incentives (i.e. rewarding consumers who forgo an air conditioner or install integrated storage with PV systems), but here again, such a model would likely appear obscure to most consumers who simply want reliable and affordable power, and inadvertently create a disincentive to reduce annual energy consumption.

Interestingly, the more obvious simple solution of encouraging the use of evaporative coolers in Victoria is bypassed in favour of a suite of more complex policy responses, including smart grids, demand management, TOU pricing, and an expansion of the energy efficiency bureaucracy (see Chap. 5). Taking the potential of smart-grids even further and trying to align industrial and consumer demand with the availability of renewable energy in near real time would represent a level of complexity at least an order-of-magnitude greater again.

These provide valuable examples of Tainter's complexity spiral: the response to the problem is increased complexity, but the complexity and the associated unintended consequences introduce a further suite of problems that need to be solved.

3.14 Energy Efficiency

Energy efficiency, voluntary frugality, clean production, waste reduction, and recycling form a standard set of policy prescriptions to reduce emissions and the depletion of resources (Alcott 2008). Such is the power and intuitive appeal of the idea of energy efficiency that it has been almost universally adopted as a key plank of the "sustainability project" by environmental NGOs, green parties, and large sections of government (Palmer 2012). Indeed, nearly all of the "100 % renewable" plans assume significant gains in energy efficiency in response to an expansion of regulatory and cost-based incentives.

The quintessential example of Jevon's Paradox was that the steadily improving efficiency of coal-fired steam engines drove an expansion of steam power. Jevon's paradox or the efficiency rebound effect has been extensively debated within the literature and remains strongly contested (Alcott 2005, 2009; Gavankar and Geyer 2010; Maxwell et al. 2011; and Sorrell 2010).

The so-called IPAT identity incorporates energy efficiency and provides a useful concept for discussing the drivers of emissions (International Panel on Climate Change (IPCC) 2000):

$$\text{Impact} = \text{Population} \times \text{Affluence} \times \text{Technology}$$

A variation on the identity is referred to as the Kaya identity, which links CO_2 emissions with affluence (GDP/capita), energy efficiency (Energy/GDP), and emission intensity of energy (CO_2/Energy) by the following equation:

$$CO_2 \text{ emissions} = \text{Population} \times \left(\frac{\text{GDP}}{\text{capita}} \right) \times \left(\frac{\text{Energy}}{\text{GDP}} \right) \times \left(\frac{CO_2}{\text{Energy}} \right)$$

The empirical evidence to date has been that energy efficiency measured as a ratio of energy consumption to global GDP has shown a steady improvement, yet emissions continue to rise—a focus on national figures can be misleading because of the outsourcing of energy-intensive industry and manufacturing. Put simply, efficiency helps us grow richer, and this in turn becomes a driver of increased resource use and population. Similarly, efficiency programs with a low transaction-cost, such as those modeled on the "Top Runner" concept, may provide broader benefits by consolidating readily available gains.

It is important here to differentiate between demand-side technical efficiency, such as vehicle fuel economy CAFE standards, and supply-driven conservation and efficiency. Given supply constraints or rising fuel costs as a proportion of income, energy consumption would be expected to decline. But in their absence, technical efficiency gains permit the "saved" income to purchase other goods and services.

Indeed, as Alcott (2012) draws attention to more generally, despite the relative shift from an Australian economy based on primary production and manufacturing to one based more on services, it is not obvious that this sectorial shift has been accompanied by a commensurate decline in energy consumption (the so-called decoupling effect) as some early analyses may have suggested (e.g. Costanza (1980)).

Fig. 3.3 Drivers of anthropogenic emissions based on Kaya identity. *Source* Based on (Raupach et al. 2007)

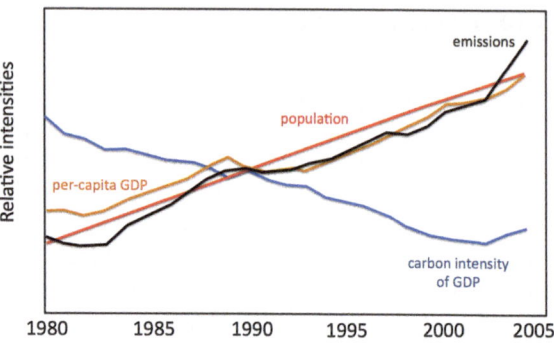

These results cast doubt on the sort of heroic assumptions often employed by the developers of "energy revolutions" described earlier in the chapter. The issue here is that an energy efficiency industry and bureaucracy is developed with laudable goals to resolve a perceived problem. But this in turn adds to the national regulatory overhead and creates additional complexity without necessarily resolving the original problem (Fig. 3.3).

3.15 Case Study: Micro-case of Rebound, the Efficiency of Melbourne's Household Space Heating

The thermal efficiency of Melbourne's housing stock has shown a significant improvement since the 1990s, and the efficiency of gas heating appliances has been growing steadily since the 1960s such that a recently constructed six-star home in Melbourne typically needs only 6 % of the energy as a typical home circa 1960 to maintain a square metre of living space at a given temperature (Palmer 2012). These improvements are a result of the evolution of technology, along with mandated standards, building regulations, and government incentives.

Yet despite these gains, the per-capita energy consumption of Melbourne's household space heating has remained remarkably stable since 1960. This counter-intuitive conclusion can be attributed to a combination of rebound and lifestyle factors including larger homes and larger heated areas, lower per-household occupancy rates, higher expectations of comfort, and an increase in the relative affordability of energy. At a micro-scale, it is apparent that notions of comfort, sufficiency, and lifestyle are bound up within the interactions between people, energy, appliances, buildings, affordability, and social values.

3.16 Case Study: Denmark's High Wind Power Penetration

Denmark is a small country with a population of 5.6 million people, with a peak demand of 6.3 GW. Its electricity system is connected with Norway, Sweden, and Germany, with a total inter-connector capacity of 5.5 GW, an unusually large capacity relative to average demand. Given that Norway's electricity system is nearly all hydro and Sweden's is half hydro, Denmark has access to the rapid ramping and start-up capability of hydro.

Hydro is highly suited to supporting the variability of wind, essentially saving Norway and Sweden's water for later use during excess wind generation and drawing down dams during calm periods. Given that there is a strong correlation between excess wind generation and Denmark's exports, there is an argument that Denmark's actual use of wind is lower than the gross generation would suggest (Green 2012). Since Norway and Sweden are usually able to purchase Denmark's excess wind generation at a lower price than the usual price when Denmark must import, there is an incentive for Norway and Sweden to facilitate the cross-border flows, with Denmark bearing the net cost of trades. Regardless, the Nord Pool spot market clears electricity trades across borders and balances supply.

Denmark provides a unique example of a small country as part of a large grid with an excellent wind resource blowing from the North Sea and unique access to large hydro resources in adjoining countries for grid balancing. Australia has limited topographic relief, with most relief lying along the eastern Great Dividing Range, which has already been exploited, most notably in the Snowy Hydro Scheme. Hence, there are few if any broader lessons for Australia from Denmark's unique experience.

3.17 Case Study: Germany's High Solar PV Penetration

In the case of Germany, PV made up 5.8 % of Germany's electricity production with 32,400 MW of PV (as at 31 December 2012). As with Denmark's wind, the instantaneous penetration of PV can appear very high within Germany, but is absorbed by the larger European grid. Figure 3.4 shows a sunny week during summer, demonstrating that exports are highly correlated with solar PV output, with effectively up

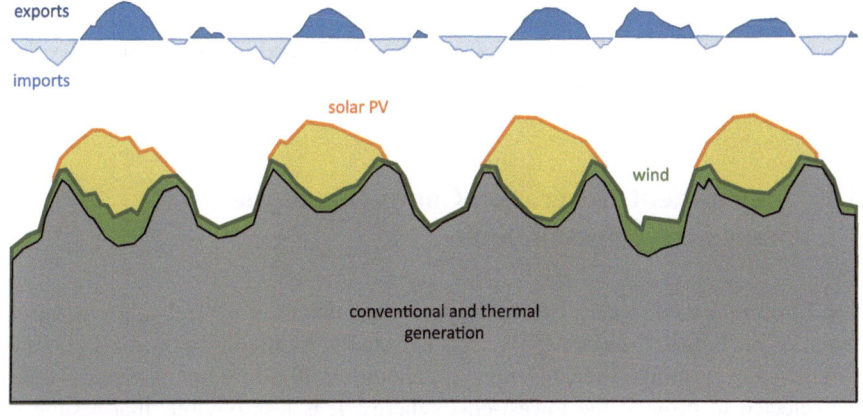

Fig. 3.4 Germany electricity production, week 31, 2012. *Source* Burger (2013)

to around 40 % of the solar power representing excess generation. But during the winter season when German demand peaks, solar output can be negligible for weeks.

The Continental European synchronous grid spans around 17° of longitude (equivalent to a separation of local solar noon of 66 min); hence, there is limited capacity for dispersed European solar to balance output across time zones. In Australia, the three largest cities, and around 90 % of the NEM demand lies within a narrower 9° of longitude (equivalent to a separation of local solar noon of 36 min), severely limiting the potential for balancing solar's diurnal cycle across NEM regions.

Although German renewable policy projects a significant expansion of PV generation, it is clear that the European grid will absorb much of the summer output; hence, without large-scale storage the Europe-wide solar penetration will remain well below the German penetration. In any event, as the wealthiest nation in Europe, it is difficult to ascertain how the German experience would translate across to the rest of the world; the cumulative feed-in-tariff obligation for German PV systems installed between April 2000 and the end of 2011 is estimated at €100 billion (USD $129 B) (Frondel et al. 2010).

Despite Spain following a similar path to Germany, the burden of the large investment in solar without an accompanying tangible benefit has weighed heavily on Spanish industry, consumers, and solar investors.

3.18   Intermittent Generation Locking in Path Dependence

Most of the high-penetration renewable plans encourage the rapid immediate take-up of wind and solar regardless of the status of carbon or other energy policies. The shift to intermittent plant as a larger part of the generation mix may force a path dependence and evolve into a suboptimal generation mix into the future (Ekins et al. 2011). This applies particular to the increasing role of gas-fired generation, which provides the rapid ramping capabilities required to support wind and solar. This may result in difficult and expensive choices in the future when deeper cuts are called for, but where path dependence has locked in a generation mix that needs to be highly flexible to accommodate renewables (Wood 2012).

3.19   Intermittent Generation Competing in the Same Low-Emission Space

The three renewable energy technologies often taken as natural allies within a "suite of renewables" are solar PV, concentrated solar thermal power (CST), and wind. However, unlike the conventional generation mix, in which generator types are selected to provide the lowest-cost synergy, it is less obvious that a suite of renewables provides a similar interaction.

Rather than providing a specific role, the particular renewable technologies are selected as much on the idea that a "suite of renewables" can provide reliable power. To the extent that costs are considered, they are usually based on the cost of capacity ($ per kW) or the levelized cost of electricity (LCOE), which represents the per kWh cost of building and operating a generating plant over an assumed financial life and duty cycle (US Energy Information Administration (EIA) 2012). When taken is isolation and excluding the large differences in capacity factor, both the capacity cost and LOCE of wind and solar can appear favourable in comparison to conventional generation, but direct comparisons between intermittent and dispatchable generation can be misleading.

For example, since intermittent plant does not directly displace conventional generation, a more meaningful comparison is to compare the LOCE of intermittent plant with the variable operating and fuel cost of dispatchable generation. Before factoring in a carbon price (see Chap. 6), these typically comprise 10–25 % of the overall LCOE of baseload generation, or around half of the cost of open-cycle gas (OCGT) [see Tables 10.3 to 10.13 (Electric Power Research Institute 2010a)].

Another method would be to add sufficient storage and capacity to provide a system with an equivalent annual availability factor to conventional generation; this would allow a more direct comparison (Palmer 2013). If the calculation is undertaken for solar PV, the initial capital cost shows at least a fourfold increase for the complete PV system (see Chap. 5).

But there is a further issue; various renewable generators will be competing in the same market space. Australian generators have traditionally earned a substantial proportion of their annual profit during the summer season when wholesale prices peak. For example in Victoria during the summer of 2010, there were 13 non-contiguous where the spot price peaked at between $1.72 and $10.00 per kWh (usually $0.03–$0.06). But the impact of all forms of solar is to suppress the peak prices during summer, subsequently reducing revenue for all generators (ROAM Consulting 2012). Sandiford (2012) has already reported a significant reduction in wholesale cost in Queensland and South Australia in 2011–2012 in response to PV despite PV contributing only a modest share of system energy. Hence, solar PV and CST will be cannibalizing each other's key markets and under normal market conditions would be regarded as strong competitors.

The suppressing effect has been frequently discussed as the "merit order effect" and has been cited as evidence of the capacity to lower the cost of energy to consumers, but the same effect also reduces the viability of solar if it was exposed to the normal market. Since these renewable energy sources have been protected by market interventions, it has been less obvious that they compete in the same market space. But in the longer run, outside of the energy and economic subsidies provided by the fossil-fuelled enterprise, they will need to compete in the open market.

In the case of wind, as discussed earlier, there is no obvious high-level synergy with solar. However, the extent to which grids can accommodate intermittency will be tested over the next 10–20 years, with some forms of intermittent plant crowding out others.

3.20 Reduction in Labour Productivity

One of the frequent claims for high-penetration renewable plans is an increase in net employment [for example Beyond Zero Emissions, page 108 (Wright and Hearps 2010)]. These claims are interesting because they are actually arguing for a substantial *decrease* in the labour productivity of the energy supply industry. The key here is that society is not getting anything it did not already have, excepting a relative improvement in emission intensity of the economy. To the extent that Australians are willing to bear the cost of reducing the emission intensity within a global context, that is better reflected in a carbon price (see Chap. 6).

As Michaels and Murphy (2009) note, the productivity of agriculture has improved steadily over the last century, with energy and technology substituting for labour. The improvement in labour productivity reduces the costs of agricultural products, which frees spending for other goods and services, and allows the non-agricultural sectors to develop. Indeed, agriculture now accounts for only 2.2 % of the Australian economy, down from 23 % in 1860 and 11 % in 1960.

If we were to contemplate encouraging more workers back into agriculture by intentionally forcing a decline in labour productivity (such as banning combine harvesters), the net result would be higher food costs for everyone and a contraction of other sectors. Similarly, an improvement in the productivity of the energy supply industry has been a primary driver of economic development, and an intentional reversal of this trend would lead to a contraction in other industry sectors, a reduction in economic activity, and therefore a decline in the tax receipts that fund the universal healthcare, education, welfare payments, and other public services.

References

Alcott B. Jevons' paradox. Ecol Econ. 2005;54(1):9–21.

Alcott B. The sufficiency strategy: would rich-world frugality lower environmental impact? Ecol Econ. 2008;64(4):770–86.

Alcott B. Impact caps: why population, affluence and technology strategies should be abandoned. J Cleaner Prod. 2009;18:552–60.

Alcott B. Mill's scissors: structural change and the natural-resource inputs to labour. J Cleaner Prod. 2012;21(1):83–92.

Anderson G. Dynamics and control of electric power systems. Zurich: Swiss Federal Institute of Technology; 2012.

Archer CL, Jacobson MZ. Supplying baseload power and reducing transmission requirements by interconnecting wind farms. J Appl Meteorol Climatol. 2007;46(11):1701–17.

Australian Energy Market Operator (AEMO). 100 percent renewables study: modelling outcomes. Melbourne: 2013.

Australian Energy Market Operator (AEMO). Technical guide to the wholesale market: 000-0264. Melbourne: AEMO; 2010.

Australian Productivity Commission. Electricity network regulatory frameworks report. Canberra: Productivity Commission; 2013.

Bindner H. Power control for wind turbines in weak grids: concepts development. Roskilde: Risø; 1999.

Blackburn TR. Distribution transformers: proposal to increase MEPS levels. Sydney: Prepared for Energy Efficiency Program; 2007.

Bower W, Gonzalez S, Akhil A, Kuszmaul S, Sena-Henderson L, David C, Reedy R. Solar energy grid integration systems: final report of the Florida Solar Energy Center Team. Albuquerque: Sandia National Laboratories; 2012.

Budischak C, Sewell D, Thomson H, Mach L, Veron DE, Kempton W. Cost-minimized combinations of wind power, solar power and electrochemical storage, powering the grid up to 99.9 % of the time. J Power Sources. 2012.

Burger B. Electricity production from solar and wind in Germany in 2012. Freiburg: Fraunhofer Institute for Solar Energy Systems ISE; 2013.

Cleveland CJ, Saundry P. Ten fundamental principles of net energy. Environ Inf Coalition. 2007.

Consulting ROAM. Solar generation Australian market modelling. Brisbane: ROAM; 2012.

Costanza R. Embodied energy and economic valuation. Science. 1980;210(4475):1219–24.

Crist P. Electric vehicles revisited: costs, subsidies and prospects. Discussion paper No. 2012-03; 2012.

Delucchi MA, Jacobson MZ. Providing all global energy with wind, water, and solar power, part II: reliability, system and transmission costs, and policies. Energy Policy. 2011;39(3):1170–90.

Diesendorf M. Greenhouse solutions with sustainable energy. Sydney: UNSW Press; 2007a.

Diesendorf M. The base load fallacy. Energy Sci Briefing Pap. 2007b; 16.

Eadie L, Elliott C. Going solar: renewing Australia's electricity options. 2013.

Ekins P, Kesicki F, Smith AZP. Marginal abatement cost curves: a call for caution. London: University College London; 2011.

Electric Power Research Institute. Australian electricity generation technology costs: reference case 2010. Canberra: Australian Department of Resources, Energy and Tourism; 2010.

Electricity Supply Industry Expert Panel. Technical parameters of the Tasmanian electricity supply system: information paper. Hobart: Tasmanian Government; 2011.

Elliston B, Diesendorf M, MacGill I. Simulations of scenarios with 100 % renewable electricity in the Australian National Electricity Market. Energy Policy. 2012;45:606–13.

Frondel M, Ritter N, Schmidt CM, Vance C. Economic impacts from the promotion of renewable energy technologies: the German experience. Energy Policy. 2010;38(8):4048–56.

Garnaut R. The Garnaut climate change review: final report. Canberra: Department of Climate Change and Energy Efficiency; 2008.

Gavankar S, Geyer R. The rebound effect: state of the debate and implications for energy efficiency research. Bren School Environ Sci Manage. 2010.

Green R. How Denmark manages its wind power. IAEE Energy Forum, 3rd Quarter. 2012: 9–11.

Halliday C, Urquart D. Voltage and equipment standards misalignment. Canberra: The Electric Energy Society of Australia; 2011.

Inhaber H. Why wind power does not deliver the expected emissions reductions. Renew Sustain Energy Rev. 2011;15(6):2557–62.

Institute for Energy Research. Levelized cost of new electricity generating technologies. 2012. Available online: http://www.instituteforenergyresearch.org/2011/02/01/levelized-cost-of-ne w-electricity-generating-technologies/. Accessed 1 Jan 2013.

International Energy Agency (IEA). Key world energy statistics: 2012. Paris: IEA; 2012.

International Energy Agency (IEA). Technology roadmap: smart grids. Paris: IEA; 2011.

International Panel on Climate Change (IPCC). IPCC special report on emissions scenarios for COP 6, part 3.1. Introduction. GRID-Arendal. 2000. Available online: http://www.grida.no/ publications/other/ipcc%5Fsr/?src=/climate/ipcc/emission/050.htm. Accessed 1 Jan 2013.

Kaffine DT, McBee BJ, Lieskovsky J. Emissions savings from wind power generation: evidence from Texas, California, and the Upper Midwest. 2011.

Kemp M. Zero Carbon Britain 2030: a new energy strategy. The second report of the Zero Carbon Britain project. Powys: Centre for Alternative Technology; 2010.

Lund H, Kempton W. Integration of renewable energy into the transport and electricity sectors through V2G. Energy Policy. 2008;36(9):3578–87.

Maxwell D, Owen P, McAndrew L, Muehmel K, Neubauer A. Addressing the rebound effect. Eur Comm DG Environ. 2011.

Michaels R, Murphy RP. Green jobs: fact or fiction. Institute for Energy Research: 2009.

Palmer G. Does energy efficiency reduce emissions and peak demand? A case study of 50 years of space heating in Melbourne. Sustainability. 2012;4(7):1525–60.

Palmer G. Household solar photovoltaics: supplier of marginal abatement, or primary source of low-emission power? Sustainability. 2013;5(4):1406–42.

PE International. Life cycle CO_2e assessment of low carbon cars 2020–2030. Low Carbon Vehicle Partnership: 2012.

Raupach MR, Marland G, Ciais P, Le Quéré C, Canadell JG, Klepper G, Field CB. Global and regional drivers of accelerating CO_2 emissions. Proc Natl Acad Sci. 2007;104(24):10288–93.

Sandiford M. Who's afraid of solar PV? The conversation. 2012. Available online: http://theconv ersation.edu.au/whos-afraid-of-solar-pv-8987. Accessed 1 Jan 2013.

Sayeef S, Heslop S, Cornforth D, Moore T, Percy S, Ward JK, Berry A, Rowe D. Solar intermittency: Australia's clean energy challenge: characterising the effect of high penetration solar intermittency on Australian electricity networks. Sydney: CSIRO; 2012.

Simshauser P, Downer D. Dynamic pricing and the peak electricity load problem. Aust Econ Rev. 2012;45(3):305–24.

SKM MMA. Garnaut climate change review update 2011: advice on change in merit order of brown coal fired stations. Melbourne: SKM MMA; 2011.

Smil V. Creating the twentieth century: technical innovations of 1867–1914 and their lasting impact. New York: Oxford University Press; 2005.

Sorrell S. Energy, economic growth and environmental sustainability: five propositions. Sustainability. 2010;2:1784–809.

Sovacool BK. The intermittency of wind, solar, and renewable electricity generators: technical barrier or rhetorical excuse? Utilities Policy. 2009;17(3):288–96.

Tainter J. The collapse of complex societies. Cambridge University Press: 1990.

Tasmania Department of Infrastructure Energy and Resources. Submission to National Electricity Amendment (network support and control ancillary services) rule 2010 [ERC 0108]. Hobart: Tasmania DIER; 2010.

Teske S. Energy [R]evolution: a sustainable world energy outlook. 3rd ed. Hamburg: Greenpeace International, European Renewable Energy Council; 2010.

Trainer T. Can renewables etc. solve the greenhouse problem? The negative case. Energy Policy. 2010;38(8):4107–14.

Trainer T. A critique of Jacobson and Delucci's proposals for a world renewable energy supply. Energy Policy. 2012;44:476–81.

Trainer T. 100 % Renewable supply? Comments on the reply by Jacobson and Delucchi to the critique by Trainer. Energy Policy. 2013.

United Nations. Multidimensional issues in international electric power grid interconnections. New York: 2006.

US Energy Information Administration (EIA). Levelized cost of new generation resources in the annual energy outlook 2012. Washington, DC: EIA; 2012.

Wood T. The future of gas power: stepping stone or snare? The conversation. 2012. Available online: http://theconversation.edu.au/the-future-of-gas-power-stepping-stone-or-snare-4575. Accessed 1 Jan 2013.

Wright M, Hearps P. Australian sustainable energy: Zero carbon Australia stationary energy plan. Melbourne: Melbourne Energy Research Institute; 2010.

WWF. The energy report: 100 % renewable energy by 2050. Gland: WWF; 2011.

Chapter 4
Electricity Networks: Managing Peak Demand

4.1 Defining Congestion

The most common example of congestion is traffic congestion, when there are simply too many vehicles competing for the same road space. But congestion can also apply to many other products and services. Standing in a bank queue, postal delays during the Christmas period, or a slow water tap are all examples of a slowed or delayed output when demand exceeds available supply. In most activities, congestion is simply a consequence of the fact that it is too expensive or impractical to cater for the peak annual demand. Eight-lane roads are not constructed to cater for a rare peak period when a two-lane road suffices for most of the time, and even if the road were widened, other connecting roads would likely congest sooner.

Figure 4.1 is a stylized graph depicting a number of products and services on a scale of relative congestion versus available storage. "Relative congestion" is loosely defined as the delay that might be considered acceptable, while the product or service remains functional; telephone congestion may require several minutes of delay before accessing a line or sending a text message on the cellular phone network, heavy traffic on roads may cause a 30-min trip delay, while the post may be delayed for a day. The adequacy of the product or service to meet acceptable performance standards during periods of peak demand could be roughly taken as the vector of both the capability to congest (while remaining functional) and the capability of prestoring the product.

Some services can utilize storage, while also tolerating mild congestion, such as the city water supply. The water supply could still function with moderate congestion, delivering water at a lower pressure and flow rate, and storage in dams in Australia is typically 2 years of consumption. Mains-delivered natural gas is also resistant to a moderate pressure loss in the mains, and the storage within pipelines may supply typically 1–5 days of demand with the loss of gas injections. In the case of the passenger rail network, patrons can be "stored" for a brief period at no cost on railway platforms prior to the train arrival, with a rapid embarkation once the train arrives, and a high-density passenger throughput within a narrow rail

G. Palmer, *Energy in Australia*, Energy Analysis,
DOI: 10.1007/978-3-319-02940-5_4, © Graham Palmer 2014

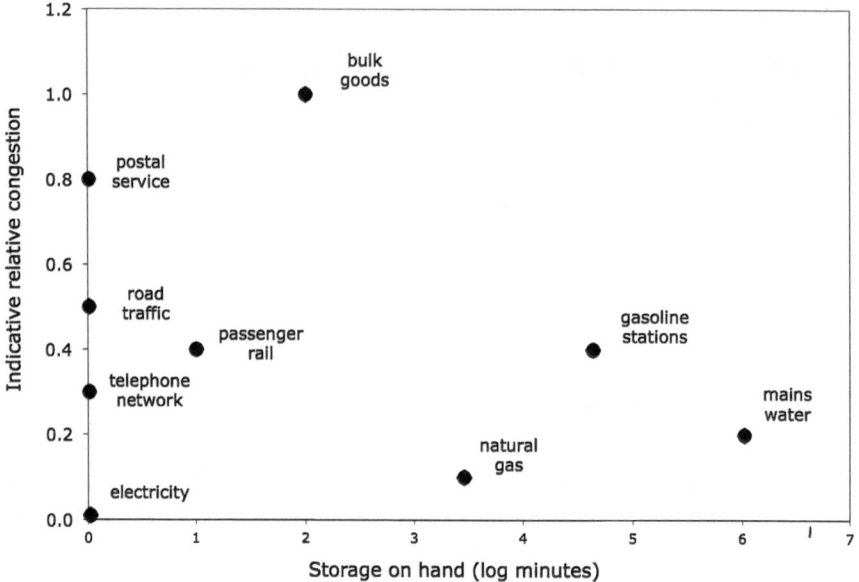

Fig. 4.1 Stylized graph of congestion versus storage

corridor. Further, patrons will usually tolerate a moderate delay if the train network is busy, allowing congestion to buffer peak demand without the rail system failing.

On the other hand, the postal service and telephone network cannot prestore services, but can operate with moderate congestion; postal deliveries may be delayed for a couple of days during the Christmas period, or there might be several minutes of delay in accessing an available channel on the cellular phone network at the end of a concert or football match.

Congestion usually provides valuable feedback to consumers to voluntarily shift demand at no cost to the product or service provider; road users will try to avoid busy freeways during peak periods, and customers are "encouraged" to use online banking to avoid bank queues.

4.2 Congestion in Electricity Networks

But the electricity network is unique in being completely intolerant to congestion and possesses no *inherent* storage within the system; hence, electricity must be consumed the instant it is produced. Further, in the absence of explicit time-of-use tariffs, it provides no feedback to consumers that congestion is imminent. Electricity is a precise product and must meet minimum standards of quality for voltage, frequency, and harmonics to ensure stability of the network, safety, reliability, longevity, and correct operation of connected appliances, motors, lighting, and other equipment.

In the event of congestion, the other utilities, including gas, water, and telephony, can remain functional without active intervention; however, the electricity network requires active intervention to shed load (i.e. blackouts). And critically, the state-imposed reliability standards are set at a very high level due to the high cost for business of blackouts and the likely political fallout if the electricity network were perceived as failing; the Australian reliability standard states that over the long term, the maximum expected regional unserved energy should be no more than 0.002 % of a region's annual energy consumption (Australian Energy Market Operator (AEMO) 2010). This has important implications for the operation of the network with an increasing penetration of intermittent generation, which possesses completely different reliability characteristics to conventional generation.

4.3 Measuring Peak Demand Events: The Load Duration Chart

The temporal profile of supply and demand can be depicted with the annualized load duration chart (see Fig. 4.2), which is plotted as the proportion of time (x-axis) in which the demand (or supply) exceeds a proportion of maximum annual demand (y-axis). The familiar S-shaped curve is nearly universal for large electricity systems, in which demand resides between approximately 45 and 75 % for most annual hours, but where there are relatively few hours at the extremities

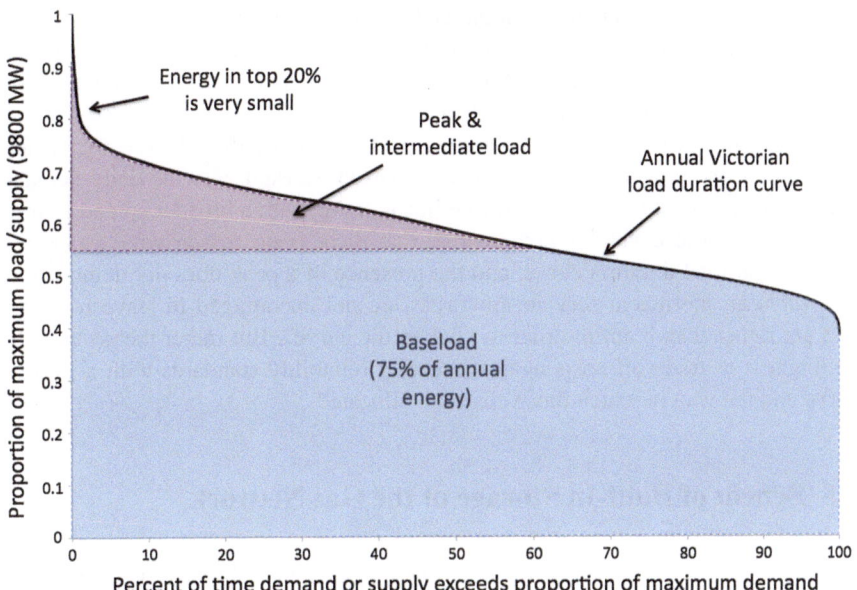

Fig. 4.2 Load duration chart for Victoria for 2010

of demand. In particular, the sharp upward turn on the left part of the graph represents the relatively few hours in which demand spikes. In Victoria, peak demand has been growing at a greater rate than annual energy demand, mostly in response to increasing penetration of air conditioning; the highest 20 % of demand is used only 1 % of annual hours; hence, peak demand is essentially an economic and equity issue, rather than a greenhouse issue. Note that the area under the graph represents the annual energy. The load duration chart provides a graphic display of the proportion of time the network is near peak demand.

4.4 Comparison of Load Duration Curve with Other Products/Services

Much of the attention on peak demand in recent years has been on the growing peakiness of the load duration curve (i.e. the sharp upward turn on the left side of the load duration chart). The issues of efficiency naturally arise as to the consequences of the load, but it is not obvious that the utilization of the electricity network is worse than many other products and services. For example, hospital emergency departments respond to a very wide range of hourly demand, but communities expect their hospitals to provide the facilities and services nonetheless. Indeed, very few activities exhibit a flat load duration chart and there is no particular reason why we should expect electricity demand to behave any differently. These types of regularly diurnal cycles are the result of normal human behaviour; we wake up, go to work or school, and return home. These cycles are reflected in heating and cooling loads, traffic and pubic transport loads, and other commercial and industrial activity.

Figure 4.3 compares the load duration curve for three services; the Victorian electricity and natural gas network for the year 2010, and a busy road intersection in Melbourne, measured over a month at hourly intervals.

The gas network shows much stronger peak behaviour than the electricity network due to the seasonality of the heating season overlaid with the daily demand cycles. The road network shows a more linear curve, with a broader spread of traffic density. These examples demonstrate that there is no reason why we should expect a flat load duration curve, and the presence of a peak does not demonstrate *ipso facto* an inefficient service; motorists are not encouraged to leave for work at 3 am rather than 8 am in order to "flatten the curve". But rather the issue is the willingness to trade-off costs against meeting reliability standards with a peakier curve and the way in which those costs are allocated.

4.5 Benefit of Built-In Storage of the Gas Network

Although the Victorian gas network is much peakier than the electricity network, the built-in storage characteristics of the network and the capability of operating at varying pipeline pressures insulate the network from congestion costs. Most of the gas injections come from the Longford gas processing plant, which draws gas

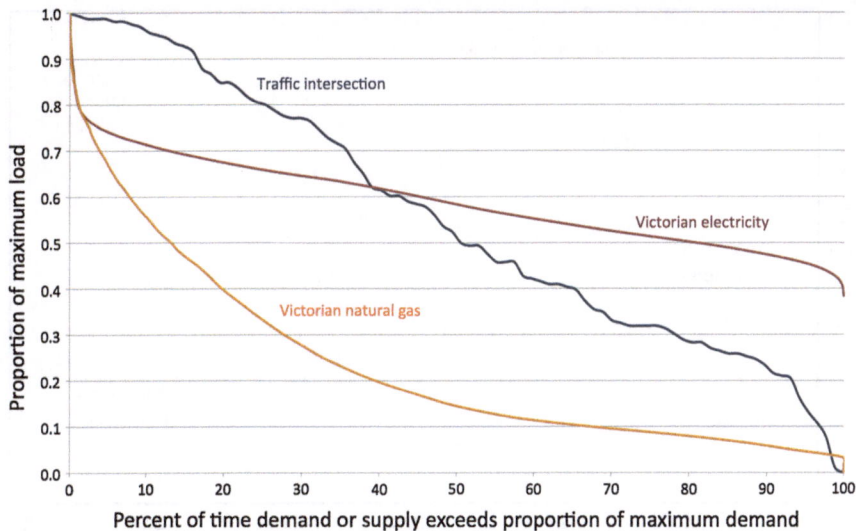

Fig. 4.3 Load duration chart for the Victorian electricity and gas networks, and a busy traffic intersection

from the Gippsland Basin gas fields. Longford only needs to operate at around half the peak demand during the winter peak period since the pipelines act as a storage medium.

The two daily peaks in demand during winter are mostly due to householders operating gas heating systems in the early morning and evening after people arrive home. Indeed, in terms of the rate of energy consumption, the annual gas peak demand for the Victorian gas network is around double that of the Victorian electricity network. This would have serious implications in the event of a large-scale changeover to reverse-cycle electric heating and would shift the annual peak to winter in the southern states (Fig. 4.4).

Given the clear benefits of the gas network, there have been proposals to use renewable sourced methane or hydrogen in the gas network; however the high cost of conversion and an ideal round-trip conversion efficiency of 36 % (Sterner 2010) will significantly add to the energy overhead and ultimately constrain its potential.

4.6 Availability Factor to Meet Reliability Standards

A motor car might only be driven a few hours a week, but it is nonetheless expected that it will start and run when required. In much the same way, the full-generation fleet will not operate at full capacity all the time, but it is expected that individual generators can be relied upon to run when required to achieve a high aggregate reliability of the fleet, even if it is only a few days a year.

The common measure of the usage of a generator is the capacity factor, which is defined as the percentage of electricity generated relative to what it

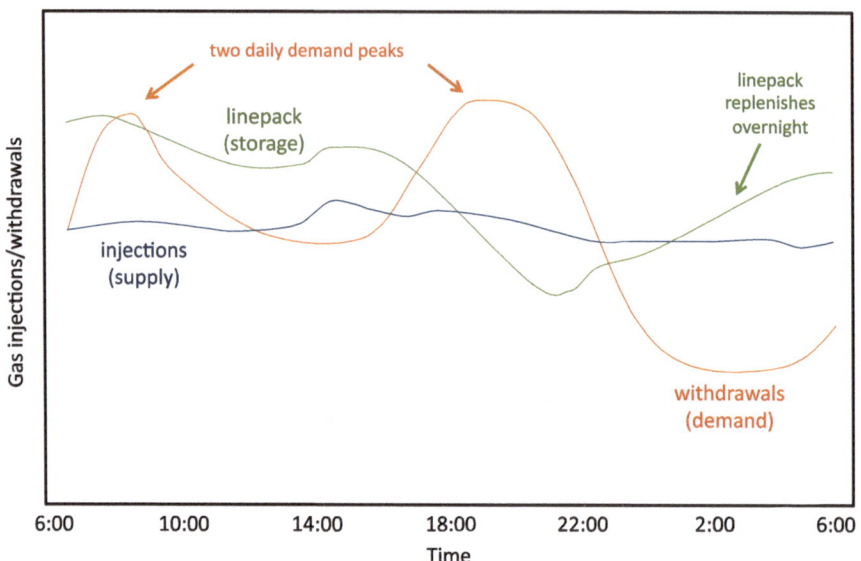

Fig. 4.4 Victorian gas network withdrawals, injections, and linepack during a winter peak. *Source* AEMO (2010)

would generate if operated at full capacity for 24 h a day, every day of the year. In contrast, the equivalent availability factor (EAF) defines the availability of a generator.

For example, the average Australia motor car that is driven 10,000–15,000 km per year may have a usage capacity factor of around 10 %, but an availability of greater than 95 %, due to the very low risk of a breakdown. In much the same way as motor cars have regular servicing scheduled at convenient times, planned maintenance for generators is typically scheduled around off-peak periods and low seasons, in coop-eration with the system operator. The average forced outage rate (FOR) in advanced countries is typically 4 % to 6 % for conventional generation, with an EAF of around 90 %. In contrast, renewable energy exhibits an availability factor that is not related to its technical reliability, but the temporal availability of the natural resource. For example, wind power may have a capacity factor of 30 %, but an availability factor of under 10 %, meaning that less than 10 % of the nameplate capacity can be relied upon to meet demand when needed. The terms "capacity credit" and "availability" are broadly interchangeable in relation to renewable energy.

4.7 Availability Versus Capacity Factor

Figure 4.5 depicts the availability factor versus capacity factor for a range of conventional and renewable energy generators based on a range of technolo-gies assessed by a recent Australian report. There are broadly 4 families of generation:

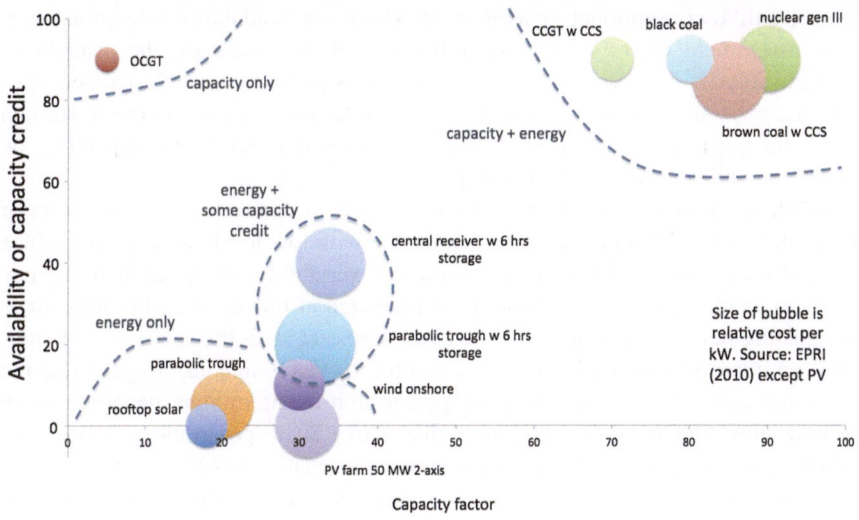

Fig. 4.5 Availability factor versus capacity factor for a range of power generators

1. Baseload and intermediate load, which provides the bulk of system energy and firm capacity. In Australia, these are coal, but may also be nuclear, or hydro in countries with sufficient hydro resources. Geothermal and biomass can also potentially provide the same role.
2. Peak load generation, such as OCGT, or in some cases hydro, which provides firm capacity but may deliver only a small proportion of total energy over a year. In Australia, hydro is used primarily in a high-value load following and peak demand role.
3. Traditional renewables, such as solar PV, wind, and tidal, which supply energy, but provide only marginal firm capacity.
4. Concentrated solar thermal with storage, which supplies energy, but depending on configuration can also contribute moderate firm capacity.

The availability factor of conventional generation is a function of the failure risk profile; well-maintained machines will have a low risk of failure, regardless of age. In Victoria, the Yallourn coal-fired generator for example, which was commissioned in 1973, had a FOR of 3.9 % and an EAF of 89.5 % in 2010 (TRU Energy 2011). Since the failure risk profile of a fleet of conventional generation is mostly uncorrelated, each additional generator adds to the aggregate reliability. The risk associated with "credible contingency events", such as the sudden loss of a large generator unit, is addressed through setting appropriate reserve margins and ancillary services.

In contrast, the "capacity credit" of a fleet of renewable generators is measured quite differently. The intermittency of wind and solar is discussed in terms of the probabilistic measure of loss of load probability (LOLP), which is a measure of the probability of wind or solar being available at times of peak demand, and is therefore very sensitive to the correlation of the wind or solar resource, respectively.

Hence, unlike conventional generation in which the availability is a straightforward assessment of the engineered reliability of the machines, the availability of renewables is both a technical reliability measure and a statistical function of both the generation profile and the demand profile. For example, in the Australian NEM, the capacity credit for wind ranges between 0.4 and 9.2 %, depending on the season and the state (AEMO 2012).

Much of the discussion about the contribution of renewables and the costs of intermittency has been based on the experience with low-level intermittency. Since the grid is already configured to accommodate variability, low-level intermittency can be treated as negative load and accommodated at low or no additional cost to the existing fleet. For example, the UKERC (Gross et al. 2006) review of the literature on intermittency, focusing on wind in the United Kingdom, suggests that the consumer costs of variability at a low penetration of intermittent penetration can be quite low. However, they also note that above 20 % penetration, "more radical changes would be needed in order to accommodate renewables". The statistical measurement of LOLP also leads to a relative decline in the capacity credit of intermittent power sources as the penetration increases (E.ON Netz 2005).

There appears to be a significant divergence in the reported costs of increased intermittency. This is likely due to a number of factors including differences in generation mix, the temporal availability of the natural resource, differing technical standards, uncertainties, and underlying assumptions. These are all issues requiring resolution. Perhaps the more interesting observation is that most of these cost-of-intermittency assessments assume the availability of a highly reliable and adaptable grid based mostly on conventional generation.

4.8 Air Conditioning Driving Rising Peak Demand

The primary driver of peak demand in Australia is increasing penetration of household air conditioning, which along with inefficiencies in the networks industry and regulatory flaws, has driven a significant increase in electricity costs (Australian Productivity Commission 2013). National penetration of refrigerated air conditioners was estimated at around 20 % in 1994, but has risen to around 60 % (Saman et al. 2009), with annual sales of around 1 million units. Evaporative coolers consume much less energy and present a load of less than 20 % of an equivalent refrigerated system, but are only effective outside of humid climates. Around 60 % of the Australian population reside in regions that are suitable for evaporative and a greater share if indirect evaporative models are considered. These used to be more popular up until the 1990s, but a number of factors have driven their relative share down significantly, despite consumers reporting a high satisfaction with their operation.

A refrigerated air conditioner will usually present the full nameplate load to the network on the hottest days, regardless of the efficiency of the dwelling or unit, since residential air conditioners will usually run at a high-duty cycle on the hottest days (an exception would be superinsulated homes). For example, the cost of supplying an additional 1 kW of power to serve air-conditioning load on hot days has been costed in one scenario at around $4,000. This suggests that in some scenarios, the capital cost of meeting the additional peak load may be several times more than the purchase price of the air conditioner.

The rising efficiency of appliances and dwellings is themselves bound up with the evolution of notions of comfort and sufficiency, which has driven a trend towards larger homes with larger heated and cooled areas, and higher demands on thermal comfort. By providing generous solar feed-in tariffs to offset greenhouse emissions and reduce energy costs, along with the promotion of "5 star" air conditioners and "5 star" homes, Australian governments have indirectly contributed to the problem since larger homes require larger air conditioners, with solar PV doing very little to reduce the annual peak load. Viewed through this alternate lens, PV and energy efficiency standards have become one of the drivers of the unconstrained expansion of suburbia and therefore as much a part of the problem; without a constraint on the overall impact, energy efficiency reduces the barriers to the evolution of comfort and "lifestyle".

4.9 Declining Productivity of Network Assets

The issue in relation to electricity networks is that a peakier load curve reduces the productive use of assets, therefore increasing costs. Since network cost recovery is based on energy consumption rather than contribution to peak demand, the cost of a deteriorating load curve is smeared across all consumers by rising tariffs, regardless of their contribution to peak load. This is equivalent to charging all motorists a per-mile charge whether they travel on congested roads or not.

The reason for tariffs based on energy is largely historical; the early meters were mechanical accumulation devices, which recorded the cumulative energy over a period of months under a static tariff, regardless of time of day. It was recognized early that an average tariff would lead to overconsumption during peak periods but under consumption in off-peak periods, thereby reducing the efficiency of the network (Simshauser and Downer 2012). However, the additional costs of running two circuits and meters to every home could not be justified, although many homes had off-peak hot water, and the SECV provided an off-peak space-heating tariff from the 1960s.

4.10 Embedded Solar PV and Resulting Peak Loads

Since solar follows the diurnal cycle, solar PV output can only occur during the day time and will often occur during peak periods. Since household solar is embedded within the low-voltage distribution network, the resulting loading on the distribution and transmission networks will be lower during periods of high solar insolation.

However, the key driver of network augmentation costs is a relatively few critical peak demand events, which, as noted above, are increasingly driven by domestic air conditioning on very hot summer days. Whereas solar supply is centred on solar noon, domestic air-conditioning loads frequently remain strong through the late afternoon and early evening as householders return home from work. In contrast, commercial air-conditioning loads reduce at the end of the work day and provide a generally better match with solar (Watt 2004). Hence, in general, embedded PV provides only marginal benefit from a distribution network provider's perspective.

The solar output profile can be shifted to slightly later in the afternoon by orientating panels towards the west, rather than the usual north (for the southern hemisphere). However, modelling shows that the resulting effect on network load is mixed (Palmer 2013). Myers et al. (2010) report similar results in a study in Wisconsin.

Under current regulatory arrangements, incentives are structured to encourage maximum annual energy production rather than provide network support—maximum annual output is derived from north-facing panels. The use of single- or twin-axis trackers will flatten and expand the output profile. However, these are unsuitable for household use and PV in built-up locations.

The issue in relation to the low-voltage distribution network is that homes with PV remain completely dependent on the network—indeed, the very fact that PV units are able to export power into the grid and earn a feed-in tariff implies that PV systems make use of the network. Further, PV consumers still make use of the billing and customer services provided by retailers. Hence, although there may be some cases in which PV provides some network support, in the main PV without storage does not provide a robust solution to rising network costs (Fig. 4.6).

Figure 4.7 graphs the Victorian electricity load duration curve shown earlier, with the resulting curve assuming that north-facing solar PV capacity equal to annual peak demand (9,858 MW) is installed in Melbourne. It can be seen that

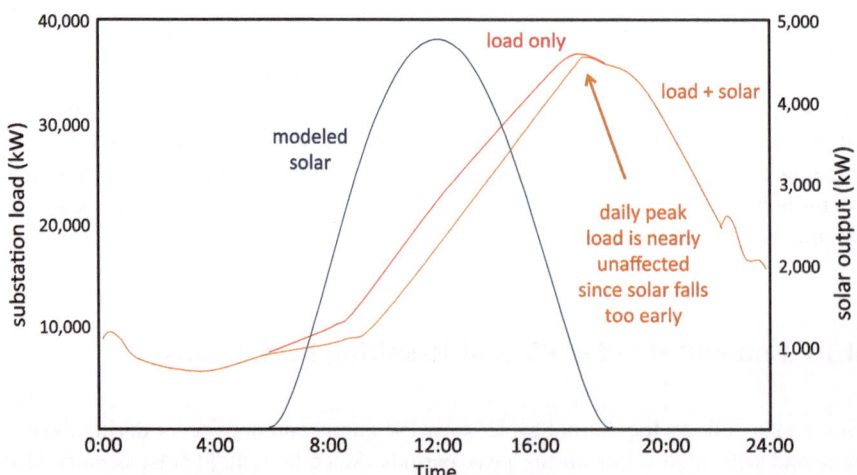

Fig. 4.6 Glenmore Park Zone substation NSW, with 6,850 customers. Solar simulation for 12 January 2010, assume 50 % have solar @ 1.5 kW. *Source* Endeavour energy

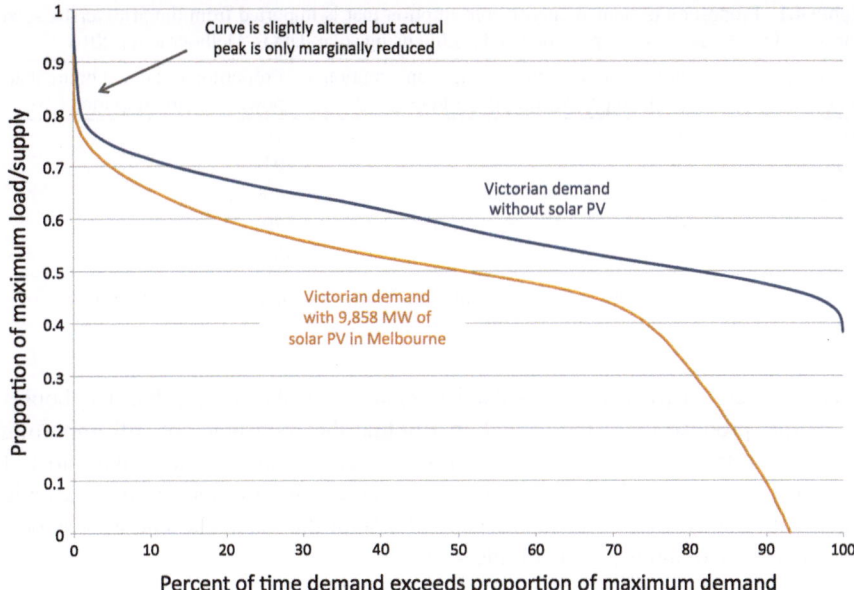

Fig. 4.7 Victorian load duration chart with 9,858 MW of solar PV in Melbourne. Based on actual demand and solar data for 2010

the embedded solar PV alters the shape of the curve slightly, but the resulting peak load is only marginally reduced, despite the solar capacity being equal to the annual peak load. The sight narrowing of the peak is due to solar insolation being coincident with the summer peak loads. But since the correlation is not robust, such as occurs on hot days with late cloud cover, or hot and humid overcast days, the resulting annual peak is only marginally reduced.

This reinforces the conclusion that solar PV does not meaningfully contribute to system capacity, but contributes to annual energy, thereby reducing fuel consumption. This has important implications discussed in Chap. 6 related to the cost of greenhouse abatement, since solar PV is not displacing generation capacity.

4.11 Implications for Household Dependence on the Electricity Network

The discussion about the contribution of solar relates solely to annual energy, but since consumers still expect electricity to be available 24 h/day, households are still completely dependent on the electricity network. To explore this, the average household demand profile was modelled using data in Deloitte (2011) with actual solar insolation for Melbourne in 2010, assuming fixed-axis North-facing solar panels, inclined at around the average roof pitch of 25°. The average Victorian household consumes 15.5 kWh/day averaged over the year, equivalent to the

Table 4.1 Proportion of annual energy consumption that is imported from the grid versus solar capacity, for a home with a typical demand profile using solar data in Melbourne for 2010

Solar capacity (kW)	Proportion of annual energy consumption that is imported from the grid (%)	Proportion of annual hours that power is being imported (%)
0.0	100	100
1.0	79	92
2.0	70	80
4.4	63	70
8.8	58	63

Assume north-facing azimuth with average tilt of 25°, corresponding to average Australian roof pitch

energy generated by a 4.4 kW solar PV system. As shown in Table 4.1, households that produce twice the annual energy that they consume are still importing power from the grid for 63 % of annual hours. The reason for this is that most of the energy is generated in a proportionally small amount of time. In Melbourne in 2010 with a fixed-axis north-facing panel, 80 % of the total solar energy was produced in 22 % of annual hours (Table 4.1).

4.12 Battery Storage to Improve PV Capacity Credit

It is generally acknowledged that solar will require built-in storage if solar penetration is to increase markedly (International Energy Agency (IEA) 2010; Sayeef et al. 2012; Wilson et al. 2010). With a moderate amount of storage (up to 4 h at full capacity), PV can provide a potentially important time-shifting or network support role and reduce the quantity of spilled energy at a high penetration of PV. According to modelling (Palmer 2013), the inclusion of 4 h of embedded storage, when operated at the optimum charge/discharge sequence with perfect hindsight, was able to reduce the peak load by 85 % of the installed solar capacity during summer (i.e. a 1 kW system would provide 850 W of assured power during peak periods).

This suggests that the most useful role for embedded solar PV is for network support. Depending on battery cost, and the costs for the specific network augmentation, the cost of storage may be competitive in a peak demand mitigation role (Electric Power Research Institute 2010).

4.13 Household PV with Storage to Improve Network Utilization

The cost of supplying an additional 1 kW of power to serve air-conditioning load on hot days has been costed in one scenario at $2,000 in distribution, $1,200 in transmission, and $800 in generation, although the cost for individual scenarios is highly dependent on the spare capacity within each element of the system.

From a distribution network service provider's (DNSP) perspective, faced with a choice between the upgrade of a durable passive device, such as a distribution transformer with a 35-year life, and the installation of an active storage system with a shorter life and with less operating experience, a rational DNSP would usually choose the more robust option unless a strong economic case could be argued. Unlike greenhouse mitigation and carbon pricing, which comprises a trade-off between emission intensity and carbon costs, network augmentation is essentially a trade-off between initial capital costs and lifetime costs.

In any event, as regulated monopolies, Australian DNSPs are able to pass through the costs regardless. Under current regulatory arrangements, it makes no sense for households to install batteries since they cannot capture the potential avoided network costs, nor access the wholesale spot market to potentially profit through arbitrage or provide ancillary services. Hence, different distribution regulatory arrangements or incentives would need to be implemented before storage will be regularly included in grid-connected household PV.

4.14 Network Support Summary

Figure 4.8 provides a comparison of the effectiveness of various solar technologies with wind to provide network support. The issue here is that generation that is embedded within the distribution network can result in deferral of network augmentation, providing that it presents a guarantee of generation during the small number of critical peak demand events. Without this guarantee, the embedded generation cannot defer the cost associated with the distribution network.

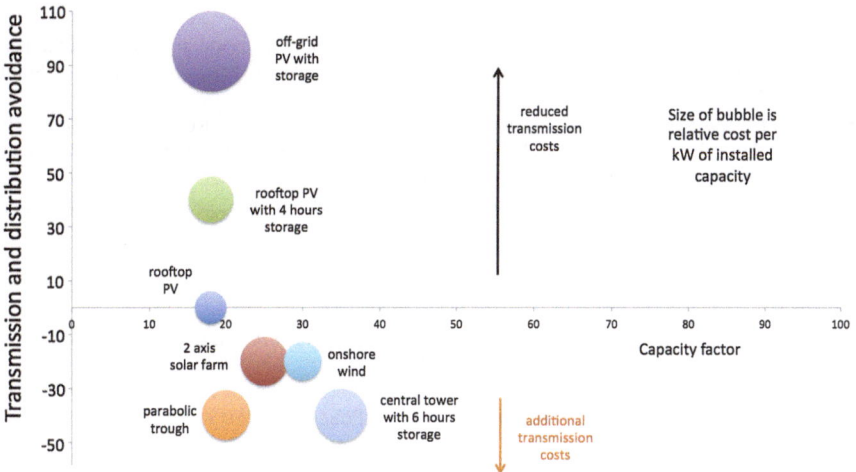

Fig. 4.8 Transmission and distribution cost avoidance versus capacity factor. Higher cost avoidance is better

Off-grid solar with storage and oversized solar capacity provides the most benefit to avoided network costs; indeed, off-grid can operate without grid connection. However, rooftop PV without storage provides essentially little deferral of network augmentation and at high penetration requires additional costs at a local distribution level due to voltage instability discussed in Chap. 3.

In the case of wind and large-scale solar, there will be additional costs associated with transmission and transformer infrastructure. CSP is particular exposed in this area since plants must be constructed in arid regions to assure sufficient annual direct sunlight, far from demand centres.

References

Australian Energy Market Operator (AEMO). Technical guide to the wholesale market. AEMO: Melbourne, Australia; 2010 (000-0264).

Australian Energy Market Operator (AEMO). *Wind contribution to peak demand*; 2012.

Australian Productivity Commission. *Electricity network regulatory frameworks report*; Productivity Commission: Canberra, Australia, 2013.

E.ON Netz. *Wind report 2005*; 2005.

Electric Power Research Institute. Electricity energy storage technology options: a white paper primer on applications, costs, and benefits. Palo Alto, CA: EPRI; 2010.

Gross R, Heptonstall P, Anderson D, Green T, Leach M, Skea J. The costs and impacts of intermittency: an assessment of the evidence on the costs and impacts of intermittent generation on the British electricity network. UK Energy Research Centre: London, UK; 2006.

International Energy Agency (IEA). Technology roadmap—solar photovoltaic energy. Paris, France: IEA; 2010.

TRU Energy. *Investor Presentation, March 2011*; TRU Energy: Melbourne, Australia, 2011.

Myers KS, Klein SA, Reindl DT. Assessment of high penetration of solar photovoltaics in Wisconsin. Energy Policy. 2010;38(11):7338–45.

Palmer, G. Household solar photovoltaics: supplier of marginal abatement, or primary source of low-emission power? Sustainability. 2013;5(4):1406–42.

Saman W, Bruno F, Liu M. Technical background research on evaporative air conditioners and feasibility of rating their water consumption. Adelaide, Australia: University of South Australia; 2009.

Sayeef S, Heslop S, Cornforth D, Moore T, Percy S, Ward JK, Berry A, Rowe D. Solar intermittency: Australia's clean energy challenge: characterising the effect of high penetration solar intermittency on Australian electricity networks. Sydney, Australia: CSIRO; 2012.

Simshauser P, Downer D. Dynamic pricing and the peak electricity load problem. Aust Econ Rev. 2012;45(3):305–24.

Sterner M. Bioenergy and renewable power methane in integrated 100 % renewable energy systems: Limiting global warming by transforming energy systems. Kassel University Press GmbH; 2010

Watt, M. The value of PV in summer peaks. *Centre for Photovoltaics Engineering University of NSW* 2004.

Wilson IAG, McGregor PG, Hall PJ. Energy storage in the UK electrical network: Estimation of the scale and review of technology options. Energy Policy. 2010;38(8):4099–106.

Chapter 5
EROI of Solar PV

5.1 Life-Cycle Assessments

Lifecycle assessments (LCAs) are the most common tool to assess the environment impacts of products and processes. LCAs provide a consistent framework to measure the inputs (energy, water, natural resources, etc.) and outputs (energy, wastes, emissions, etc.). All energy devices require energy inputs, in their initial capital, along with recurrent operational energy. In general, renewable energy technologies have the vast majority of their embodied energy invested in their manufacture and installation. In contrast, fossil-fuelled energy technologies incur a significant proportion of their lifetime embodied energy during the long operational phase.

The embodied energy of fossil-fuelled, nuclear, and renewable energy has been an active area of research over many years. The calculation of embodied energy relies on LCAs, and there are broadly three methodologies to calculate embodied energy: process analysis, input–output analysis, and hybrid analysis, which aim to combine the two base methods and thereby partially neutralize the errors of each. This discussion will concentrate on renewable energy technologies, especially solar photovoltaics.

5.2 Energy Return on Investment

A common measure for renewable energy is the payback time (i.e. the number of years it takes for the PV and wind to generate the energy it took to produce the system). A related measure, the energy return on investment (EROI), is a dimensionless term expressing the ratio of energy generated over its lifetime relative to the embodied energy and was pioneered by ecologist Charles Hall and others in the 1970s (Gupta and Hall 2011). The EROI, payback, and related analyses provide a theoretical foundation for exploring the efficacy of specific technologies or projects (Lenzen 2008). Further, some researchers argue that the use of such "biophysical"

G. Palmer, *Energy in Australia*, Energy Analysis,
DOI: 10.1007/978-3-319-02940-5_5, © Graham Palmer 2014

properties provides insights that are not otherwise available through the strictly economic metrics employed in classical economics (e.g. Cleveland et al. 1984).

Energy payback time (EPBT) is defined as

$$EPBT = \frac{Embodied\ energy}{Energy\ out_{year}}$$

and

$$EROI = \frac{Energy\ out_{lifetime}}{Embodied\ energy_{lifetime}}$$

EPBT and EROI are related by the equation

$$EROI = \frac{Projected\ life}{EPBT}$$

where EPBT and project life are given in years

5.3 LCAs are Defined in Units of Primary Energy

A complicating factor is that conventional PV LCA analyses are expressed in terms of primary energy, but since fuels have differing quality and usefulness (for example, a joule of electricity is more useful than a joule of heat from coal), there is an argument that the EROI should include some provision to account for the varying usefulness (Murphy et al. 2011). The standard use of primary energy provides a consistent framework for LCAs, but may not always deliver the most meaningful results.

For example, Raugei et al. (2012) argue that the EROI of PV should be include provision for the average electricity thermal generator efficiency (η_{grid}) to account for the fact that PV generates electricity directly, rather than via a heat engine as occurs with most generation. Taking a typical grid efficiency of around 0.31 thereby increases the "primary energy equivalent" ($EROI_{PE-eq}$) around threefold.

Indeed, in a context in which PV displaces the use of high-cost diesel in remote grids, discussed later in this chapter, the conversion can make a lot of sense. Other examples include end-uses dependent on electricity such as lighting and electronic devices. However, the validity of such conversion is highly context specific and will not apply in most cases. It assumes high fuel substitutability and conversion efficiency from electricity to other fuels (Murphy et al. 2011) and also ignores the stochasticity of PV. Since electricity only accounts for 18 % of global final consumption of energy International Energy Agency (IEA) 2012, it is not obvious that applying a universal threefold conversion factor is appropriate; indeed, the conversion can also work the other way (Prieto and Hall 2013).

For example, liquid fuels are far more valuable than electricity for transport applications, and the electricity-to-wheels conversion efficiency is usually low; studies typically report an electricity-to-wheels conversion efficiency of no better than 25 % for electricity-to-hydrogen-based transport (see Bossel 2004; Shinnar 2003).

Transport makes up around a third of global primary energy, and in the case of aircraft, shipping, heavy road, mining, and other heavy equipment, liquid fuels appear to be a necessity for the foreseeable future (Smil 2010). On the other hand, in the case of light vehicles, electric vehicles (EVs) offer expansion potential over the coming decades, but the commercial viability of EVs in the highly competitive mass auto market remains unclear for the foreseeable future.

5.4 Early EROI Studies

Early analysis of EROI focused on the declining EROI of oil production, since it became apparent that much of the so-called "easy oil" had been discovered and exploited, with new developments requiring far greater investment, both economic and energy, to produce a given quantity of oil (Cleveland 2005). Oil producers must drill deeper, operate in harsher environments, and use horizontal drilling and enhanced recovery technologies to produce the same oil that used to be readily available with much simpler drilling operations (Tainter and Patzek 2011). Partly offsetting this are improvements in efficiency and productivity of the oil supply industry. Similar analyses have been conducted on the various solar technologies, wind, nuclear, coal, and others. Unlike oil production, which has shown a long-run secular decline in EROI, many renewable technologies have shown the inverse relationship, especially solar PV and, to a lesser extent, wind.

5.5 Reliance on Process Analysis to Ascertain EROI

The most common method used for these analyses is process analysis, which requires an itemization of every component and step in the value chain, relying on databases such as "Ecoinvent". This permits a very specific analysis, but requires the setting of upstream boundaries to keep the analysis manageable. For a given boundary, the resulting analysis can be very accurate, but leads to "truncation error", in which the upstream components are excluded (Crawford 2008).

It is often assumed that the "tail" beyond the boundary represents only a small proportion of the overall energy content (these upstream processes require proportionally much greater effort relative to their significance); however, Lenzen (2000) and Treloar (1998) suggested that the tail can make up 50 % or more of the total energy content. Crawford (2008) suggested that the error may be even greater. The exclusion of the tail may not be as important for comparative assessments within a product class (Crawford 2008), where for example, various types of solar panels are being compared, but may lead to misleading conclusions where broader evaluations are being sought, often with policy implications.

For example, process analysis typically ignores the role of capital inputs, such as the machinery used for the manufacture of materials, in part due to the difficulty in data collection. However, these may make up 10–22 % of a product's embodied energy (Crawford 2008; Lenzen 2001).

5.6 Input–Output Analysis and Hybrid Analysis

An alternative approach is input–output (I–O) analysis, which uses the monetary transactions between industry sectors to compile a life-cycle inventory. The strength of I–O analysis is that it avoids the problem of boundaries and circularity, but requires conversion of money to energy units, and suffers from aggregation errors (i.e. expensive products within an industry sector may not be the most energy intensive within that sector). Hybrid analysis uses elements of both base processes to remove boundary and aggregation errors, hence exploiting the strength of both methodologies. Hybrid analysis can be further subdivided into process-based hybrid and input–output-based hybrid

For example, Crawford and Stephan (2013) found that process analysis significantly understated the lifetime embodied energy of certified passive homes compared to more comprehensive input–output-based hybrid analyses, suggesting that such schemes may not deliver the expected lifetime energy savings. Similarly, Crawford (2008) compared the results of a process analysis of a building-integrated photovoltaic (BiPV) system with the more complete hybrid analysis and found that the hybrid resulted in around a threefold greater embodied energy.

5.7 Variance in Process-Based Solar LCAs

In the case of solar panels, a further source of divergence has been the allocation of energy of the very energy intensive process of silicon purification and crystallization, since the process is shared with the electronic semiconductor industry. For many years, solar wafers used "off-grade" material as a by-product of the semiconductor industry, leading to differing judgements as to the fair allocation of energy. In recent years, solar-grade silicon has become a more dedicated material stream allowing a more deliberative allocation of energy, along with a substantial decline in embodied energy as technology and thin wafer processes have developed. Other variations include the choice of location for the renewable energy and associated environmental factors (i.e. solar insolation), lifetime, power mixes in PV wafer manufacture, and operational energy, including cleaning, maintenance, and other factors.

5.8 Conventional LCAs Ignore Intermittency

A further major unresolved problem with LCA analyses as it applies to renewable energy has been that intermittent generation does not directly substitute for conventional generation, but nonetheless can still perform a useful role in emission abatement and fuel reduction. In recent years, there has been rapid growth in wind and solar, but it is clear that this additional generation has not replaced conventional generation in the usual sense; wind and solar tend to add to the energy mix without forcing the retirement of conventional plant.

For example, there has been around a 32 % increase in Germany's generation capacity over the past 10 years, much of it wind and solar, along with a significant scaling up of transmission and distribution infrastructure, but annual consumption has barely changed (see German Federal Ministry of Economics and Technology Bundesministerium für Wirtschaft und Technologie 2013). Therefore, the cumulative embodied energy of the electricity sector is increasing, but the energy output is not. Germany is a wealthy country that can evidently afford to exercise discretion in energy policy, but most countries cannot. The implications of intermittency will be discussed later in the chapter with several case studies.

5.9 Standard Methodological Guidelines Use Narrow Boundaries

The IEA photovoltaic power systems program (IEA-PVPS) (Fthenakis et al. 2011) provides methodological guidelines for PV life-cycle assessments, which provide a coherent and consistent framework for comparing various types of PV modules. However, the IEA-PVPS recommends process-based LCA against hybrid analysis. It defines system boundaries to allow a consistent framework for comparing different PV systems, but, in defining boundaries, necessarily excludes energy inputs that are a legitimate part of PV; for example, process-based LCAs usually exclude upstream capital inputs, the boundary ends at the inverter output, and PV curtailment and storage is not considered despite it being acknowledged that these downstream elements will be essential as PV penetration increases (Denholm and Margolis 2007; Palmer 2013).

5.10 Widened Downstream Boundaries Dramatically Alter Resulting EROI

For example, Prieto and Hall (2013) conducted a comprehensive study of Spain's rapid PV expansion during 2009 and 2010 and included operation and maintenance costs, access, fencing and associated costs, financial, legal and insurance costs, and other costs that are essential but not commonly part of process analysis. Similarly, Palmer (2013) found that the inclusion of even modest storage (from 1 to 4 h) had a significant effect on the resulting EROI, while Weißbach et al. (2013) found that the cost of "buffering" using pumped hydro also had a significant effect on the system EROI. All of these studies found that the resulting EROI of

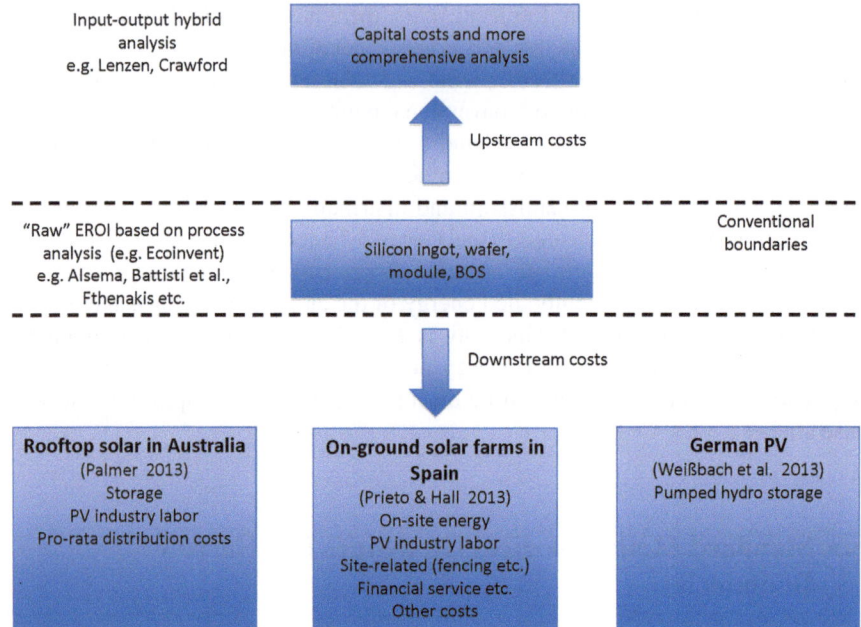

Fig. 5.1 Boundaries of photovoltaic analyses

grid-connected solar dropped to between 2 and 3 and reveals the classic problem of the weakest link in the chain; when a system approach is taken, a favourable raw EROI for solar modules is completely overwhelmed by the other components in the system (Fig. 5.1).

5.11 EROI Based on Narrow Boundaries Overstates Actual Value of PV

The problem here is not the limited boundaries per se, since well-established guidelines provide consistency when comparing products within a product class. The problem, rather, is that conclusions about the entire class of products may be drawn based on incomplete analyses.

The issue is perhaps clarified when one considers that lifetime energy returns of 15–60 based on process analyses are being quoted by authoritative sources in defence of solar PV (for example Fthenakis 2012). Yet it is not obvious that PV is producing the significant improvements in productivity, human welfare, or wealth that other energy transitions have produced, often with much lower EROI.

Even if all of the grid-connected solar PV systems in the world were to be switched off from midnight (local time), there would barely be any noticeable difference in the functioning of society and industry. Indeed, the counter-argument is much easier to sustain; for example, Spain was following Germany's lead in expanding photovoltaics and enjoyed a brief expansion in economic activity, but since 2010 has undertaken a radical revision and wind-back of its generous PV schemes as industry and consumers confront rising energy costs (Prieto and Hall 2013).

5.12 Comparison of PV Revolution with Steam and Industrial Revolution

A comparison of the current PV revolution with the eighteenth and nineteenth century steam revolution may be informative here. With the benefit of hindsight, we can see that Watt's external condenser and subsequent innovations tripped Newcomen's machine over the EROI threshold, lifting the efficiency from 0.5 to 3 % (Kümmel 2011). It is now obvious that the steam engine fuelled by coal was able to bootstrap its own energy to expand the role of steam and coal, despite the very low efficiency of the early low-pressure models. High-pressure steam eventually found multiple roles in stationary applications, rail and shipping.

Whereas the steam revolution was autonomous and self-sustaining, PV is a derivative of the fossil-fuelled industrial enterprise; there is little evidence of PV bootstrapping its own energy, powering the mines, the manufacturing facilities, transport, and the complex value chain necessary to deliver workable PV systems.

Next, steam possessed the basic qualities of all modern machines; it could be run continuously, controlled, and dispatched on command, whereas PV is neither readily controllable nor dispatchable and possesses a very low capacity factor.

The development of coal-fired steam eventually paired with the improved Bessemer mild steel process to accelerate the industrialization of the developed world in a virtuous cycle; indeed, coal, steam, and steel were defining features of modernity for much of the late nineteenth and early twentieth centuries.

Given that early steam was inefficient and almost certainly a much lower EROI than the commonly quoted EROI figures for modern PV, it is clear that the raw EROI, derived with limited boundaries and exclusive of downstream energy costs, provides little guidance of the actual value of PV to society *vis-á-vis* fossil fuels (Fig. 5.2).

5.13 Front-Loading of Embodied Energy Impairs Doubling Time

Nearly, all EROI assessments take the lifetime embodied energy as a single static number, which is used in the denominator of the EROI equation to establish the lifetime EROI. However, the timing of the embodied energy can have a significant impact on the payback timing of the energy source (King 2013).

Figure 5.3 compares two energy technologies with a lifetime EROI of 9.4 (representative conventional PV EROI from Raugei et al. 2012) and a 30-year life; one is a hypothetical fossil-fuelled generator with 20 % front-loading of embodied energy, and the other is the "raw" EROI of solar PV of the panels and BOS only. It is assumed that a third of the generated energy is ploughed back into constructing more generation.

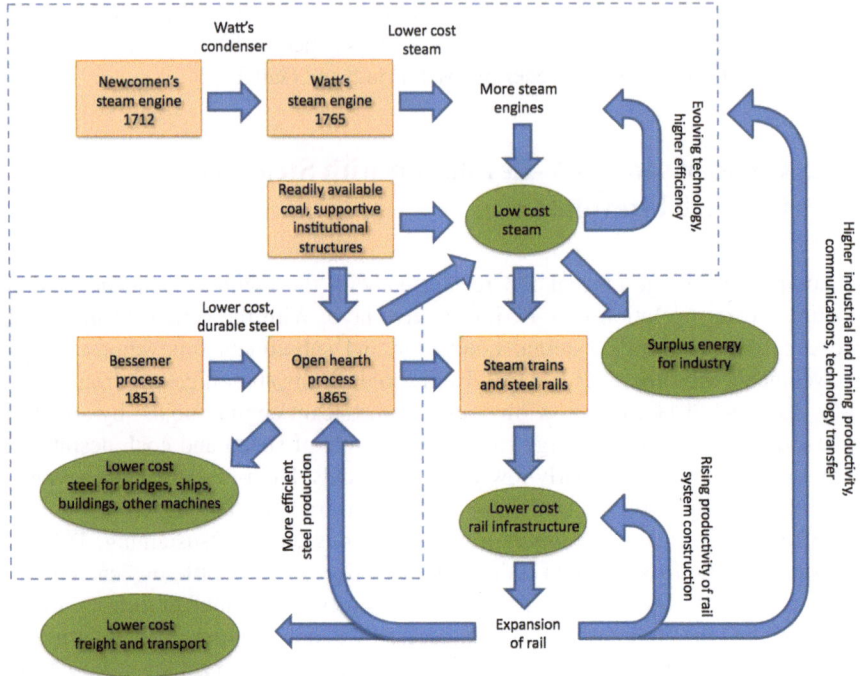

Fig. 5.2 The coal/steam/steel virtuous cycle of the industrial revolution

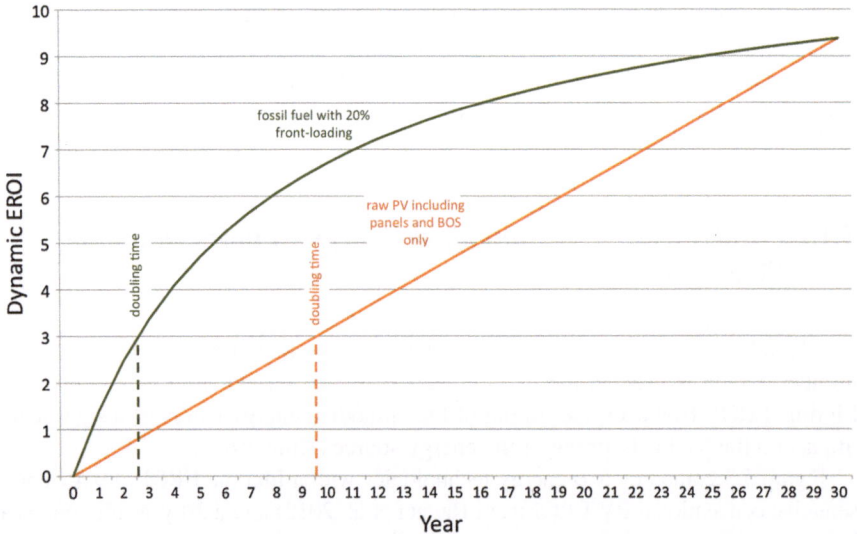

Fig. 5.3 Dynamic EROI of fossil fuel versus PV system with the same lifetime EROI

The insight is that the doubling time of the technology (assuming that a third of the energy is reinvested to produce more of the energy devices) is around three times longer than the solar PV, despite both technologies having the same lifetime EROI. The reason for this is that the solar requires the lifetime embodied energy invested as an initial upfront payment, but the fossil-fuelled generator has most of the embodied energy invested over the life of the device. This casts doubt on the sort of "solar-breeder" system postulated by some solar advocates (e.g. Blakers and Weber 2000).

5.14   Case Study: Solar PV Supplementing Coal-Fired Generation to Reduce Emissions

Consider the following hypothetical scenario, one of Victoria's large brown coal generators, which draw coal from an adjacent open-cut mine, has a large solar PV farm built near to the plant—coal makes up around 75 % of Australia's annual electricity generation. When the PV farm is generating power during the day, the coal-fired generator ramps down, thereby reducing coal consumption, then ramps back up when there is no sunshine. For the sake of simplicity, assume that there is no thermal cycling emission penalty due to the PV-induced ramping, and that the adjacent mine uses electricity directly from the generator for the mining excavator and conveyors. Given that there is around 500 years of reserves in the Victorian Latrobe Valley at current production rates, also assume that depletion of coal reserves is not an issue requiring immediate resolution. So what effect does the PV have? (Fig. 5.4).

There are two ways to look at this scenario. The first is that PV generates zero-emission energy, which is *added* to the national electricity system, but some energy is *subtracted* from another generator somewhere else, since dispatchable generation must be maintained at all times. The second is that the net-energy output of the coal/solar system is unchanged with the addition of solar, but the resulting coal consumption is lower; after all, the whole point of the solar is to displace the coal generation when there is sunshine.

In the case of Victoria's open-cut coal mining operations, the marginal cost of coal production is very low both economically and energetically. The mining operations require few staff and run mostly on electricity generated by the adjacent generators; hence, there is virtually no net-energy saving if the coal consumption rate is lowered moderately. In theory, if the excavator and conveyor plant is operated at a lower production rate, there may be a slight reduction in maintenance costs, but all of the capital, operating, and labour costs will be largely unchanged.

For simplicity, we could assume that the coal/solar system is a black box, with energy inputs and electricity sent-out. Energy inputs would include the embodied energy within the physical infrastructure, liquid fuels, energy associated with labor, and other inputs. To an outside observer who is unaware of what is going on inside the black box, the system appears to magnify the energy available to society; it takes energy provided *by* society in a number of forms, performs

Fig. 5.4 Loy Yang power station adjacent to open-cut mine. *Source* Palmer

transformations on non-renewable resources, and then delivers energy *to* society as electricity. The system EROI defines the lifetime magnification of energy, and in this case, the magnification factor would be very high. To an outside observer, a slight improvement in efficiency of the coal-fired generation would seem to deliver exactly the same energy output and emission intensity as installing solar.

Since the quantity of electricity sent-out by the coal/solar system is the same with, or without solar, but the solar incurs an energy debt in its embodied energy, the EROI of the total system must be lower than the coal-only system, and the solar component must be zero.

Let

$$EROI_{PV} = \frac{\text{Energy out}_{coal + PV} - \text{Energy out}_{coal}}{\text{Embodied energy}_{coal + PV} - \text{Embodied energy}_{coal}}$$

since

$$\text{Energy out}_{coal + PV} = \text{Energy out}_{coal}$$

and

$$\text{Embodied energy}_{coal + PV} > \text{Embodied energy}_{coal}$$

then

$$EROI_{PV} = 0$$

At first sight, this conclusion may seem counter-intuitive, but if we take a system point of view, it is obvious that the solar does not *replace* anything; it merely allows a fossil-fuelled generator to operate with slightly lower emissions. The solar is providing an abatement strategy (i.e. we are trading off emission abatement versus cost), but not an energy strategy. Hence, such an analysis should be in the economic domain and relative to other abatement strategies, discussed further in Chap. 6.

5.15 Case Study: Wind and Solar Reducing Diesel Consumption in Isolated Grid

The next example considers the case of isolated grids on islands that use diesel-fired generation, supported by wind and/or solar. The case considered here is King Island, which is located in Bass Strait in the south-east corner of Australia. King Island has a population of 1,600 and is primarily a farming community, producing beef, dairy, and seafood. It is also a quiet tourist destination from the Australian mainland.

Other similar examples in Australia include Esperance, Hopetoun, Denham, and Rottnest Island. The Flores and Graciosa Islands in the Azores, Portugal, also provides similar, highly integrated examples of wind/diesel systems in isolated small grids (see Murdock 2012 and Ilic 2011). King Island is an interesting case study for wind because it is exposed to the Roaring Forties trade winds, which bring strong westerlies.

At a cost of diesel of AUD $1.50 per litre, the cost of diesel works out at 41 cents per kWh of electricity. Operation and maintenance, depreciation and return on assets add another 17 cents per kWh, totalling around 58 cents per kWh (Government Prices Oversight Commission 2008), or around fifteen times the average wholesale cost of electricity on the mainland. The total annual cost of the system was $10.8 million in 2006/2007 (Government Prices Oversight Commission 2008).

Due to the very high cost of generation, the electricity system is highly subsidised with revenue from the island amounting to only 39 % of the actual cost. The usefulness of analysing an isolated grid with diesel is that it is easy to demonstrate an unequivocal decrease in fuel consumption as a consequence of renewable energy within a closed system (Table 5.1).

The King Island electricity system is managed by the State Government of Tasmania (through Hydro Tasmania) and has provided an opportunity to assess renewable energy. The system has five wind turbines, totalling 2.5 MW, a dual-tracking monocrystalline solar PV system rated 100 kW, a 1,500 kW resistor bank to enable excess wind to be absorbed rather than spilled, and a 200 kW, 800 kWh vanadium redox battery (since inoperable).

The system has recently been upgraded with an uninterruptible power supply class diesel engine (D-UPS) and is due for a new battery system and an enhanced demand management system. The D-UPS unit contains a large mass flywheel, which uses excess wind energy rather than diesel power to maintain its motion. Under normal circumstances, diesel/wind systems require the primary diesel to be operating at all times to maintain primary frequency control under a changing supply load. But the

Table 5.1 Summary of King Island electricity system

Annual electricity consumption	14,517 MWh
Average demand	1.7 MW
Peak demand	3.4 MW
Diesel capacity	6.0 MW
Diesel-powered generation	10,596 MWh
Diesel consumption	2,887 kL
Solar capacity	0.1 MW
Solar generation (est.)	200 MWh
Wind capacity	2.5 MW
Wind generation	3,721 MWh
Wind capacity factor (after spilling excess energy)	18 %
Diesel saved due to wind	1,068 kL

Source (Government Prices Oversight Commission 2008; Hydro Tasmania 2008)

use of a D-UPS unit permits the use of an oversized wind system, with resistor bank to act essentially as spinning reserve (when wind power exceeds usable power, the resistor bank absorbs the excess energy, and since it is operated with fast-switching electronics, the resistor load can be rapidly decreased). In the event of a rapid loss of wind, the flywheel-enabled diesel can fill in the gap before the primary diesel starts up and provides primary frequency control. The new battery will be rated at 3 MW and store 1.6 MWh, providing an estimated 30 min of supply at average load.

A meta-analysis from Kubiszewski et al. (2010) provides an average EROI of wind from operational studies of 19.8. Although King Island is very windy, much of the wind energy that is captured must be spilled or absorbed, with a net capacity factor after spilling of 18 %. A proposal to extend the wind power by 4 MW would increase the wind penetration to 65 %, but reduce the net capacity factor down to an estimated 15 %. Since most of the studies in the Kubiszewski et al. meta-analysis had a capacity factor of 30 % or greater, the actual net capacity factor probably reduces the resulting EROI to around half, or to an estimated 10 using Kubiszewski et al. as a reference.

The 100 kW of panels are mounted on two-axis trackers, which hold 24 modules and house the output, communication and control electronics. The system was projected to generate around 200 MWh of energy per annum. In practice, some of the solar power will be spilled (as happens with wind). Unlike wind generators, which possess some physical inertia, solar PV contains no inertia, and a single installation will be highly susceptible to cloud transients causing significant voltage transients in a small grid. The use of fast-ramping storage will improve PV usefulness.

However, unlike the first example in which the solar displaced inexpensive coal, in this case high-quality energy in the form of diesel fuel is displaced; hence, the solar (and wind) has a much higher value. This may provide an example that supports the use of a "primary energy equivalent" multiplier postulated by Raugei et al. discussed earlier.

King Island showcases the technical capabilities of renewable energy, storage, and highly integrated control systems, demonstrating that a high penetration of renewables is technically possible in a small grid. These systems make a lot of

Table 5.2 Comparison of national costs of electricity versus King Island

Australia—electricity national value added	
Generation	$7.0 B
Distribution	$9.5 B
Transmission	$2.2 B
On-selling and market operation	$1.4 B
Total value added 2006–2007	$20.1 B
Per capita value added	$913
Consumption	230 TWh
Per capita consumption	10.45 MWh
Average taxable income	$49,521
King Island—electricity	
Total cost	$10.8 M
Per capita cost	$6,750
Consumption	14,517 MWh
Per capita consumption	9.1 MWh
Average taxable income	$38,405

Source ABS 2012, Government Prices Oversight Commission 2008, Hydro Tasmania 2008

sense when the cost of generation is very high. From a system perspective, the displacement of high-value diesel fuel represents perhaps the highest value context that wind/solar systems can provide, but the very high cost ensures that diesel-fired generation represents only a tiny fraction of global electricity generation.

Table 5.2 provides a broad comparison of the costs of supplying electricity to Australia nationally versus the King Island supply. Although the Australian "national value added" measure is not precisely the same as the "total cost" of the electricity network on King Island, it provides a reasonable comparative measure and is consistent with the significant wholesale cost difference. The per capita cost of the electricity system on King Island is around seven times higher, and when taxable income is included, the relatively affordability (excluding subsidies) grow to a tenfold difference.

It is not surprising that the economies of scale and access to much cheaper coal and gas on the mainland provide much cheaper electricity than on King Island. It is also a reminder that the conventional electricity supply model of large centralized generators is a very efficient way of providing affordable and reliable electricity. The most fundamental problem demonstrated by King Island is that its electricity system is highly subsidised by the rest of society; King Island's export income is insufficient to cover the cost of fuel or the capital to build renewable infrastructure.

In the absence of subsidies and the bureaucratic support from the state and federal governments for public education, universal healthcare, infrastructure, and other services, it would need to resort to selling produce and tourism services in return for hard currency. This would allow it to purchase fuels for essential services, survive with a much lower energy use, and perhaps purchase more modest renewable infrastructure.

Hence, the lessons from the high-penetration renewable system on King Island are mixed. On the one hand, these schemes prove that there is no fundamental *technical* obstacle to increasing the uptake of wind and solar. But given that the King Island electricity system is highly subsidised both economically and energetically, it is difficult to draw general conclusions as it would apply at a national or global scale.

5.16 Case Study: Off-Grid Solar PV with Batteries

The third example will be of an off-grid solar PV system with batteries. The following calculations will use a recent conventional LCA review by Raugei et al. (2012), which provides an authoritative assessment for four types of cell using conventional boundaries. This discussion will use the less energy intensive (higher EROI) ribbon Si (see Table 5.4). Raugei et al. have used the average southern European insolation, which is comparable with the main population centres in the Australian NEM and a performance ratio of 0.75 (Fthenakis et al. 2011), which derates the performance to allow for cell degradation and the difference between AC inverter output, and the module's rated DC performance (Table 5.3).

Using the assumptions above, it is possible to estimate the EROI for the limit condition of an off-grid system. Indeed, off-grid solar PV is frequently used in rural Australia in contexts where the cost of connecting to the nearest feeder is sufficiently high to justify the substantial capital outlay of an off-grid solar installation.

Non-critical stand-alone systems are commonly designed with 95 % availability (equivalent to 5 % loss of load) as a compromise between cost and utility (Wenham et al. 2006)—a backup generator will be used to fill in during extended overcast periods in winter. This provides comparable availability to conventional generation and a useful insight into the limit condition of a very high penetration of PV with PV providing a quasi-baseload role.

Table 5.3 Calculation of energy return on energy investment (EROI) of PV EROI including BOS, from Table 1 (Raugei et al. 2012)

	Ribbon Si (rooftop)
Insolation [kWh/(m^2 y)]	1,700
Performance ratio	0.75
Module efficiency	13 %
$E_{out, y}$ [kWh$_{el}$/(m^2 y)]	166
T [yr]	30
E_{out} [kWh$_{el}$/(m^2)]	4,973
E_{pp} [MJ$_{PE}$/m^2]	1,907
E_{pp} [kWh$_{PE}$/m^2]	530
Solar EROI$_{el}$ = E_{out}/E_{pp} (refer Raugei et al. 2012)	9.4

Table 5.4 Calculation of EROI for off-grid solar system over 30 years

Daily energy used (kWh)	**15.5**
Solar capacity (kW)	11.1
Battery capacity (kWh)	63
Lead–acid (recycled) embodied energy (MJ/Wh) (refer p. 21 Alstone 2012)	0.87
E_{batt} (MJ) @ 4 sets over 30 years	219,240
Solar area (m^2)	79
E_{pp} (MJ_{PE}/m^2)	3,257
$E_{solar} = $ solar area $\times E_{pp}$ (MJ)	258,234
$E_{system} = E_{pp} + E_{batt}$	477,474
E_{used} @ 15.5 kWh/day over 30 years (MJ)	611,010
System $EROI_{el} = E_{used}/(E_{solar} + E_{batt})$	1.3

Excludes generator and other ancillaries

Half-hourly solar data for Melbourne, along with half-hourly household demand data from Deloitte (2011), were used in a spreadsheet model with VBA macros to model the system with storage. Deloitte's demand data provide for an average 15.5 kWh demand per day, but the magnitude of the daily demand does not alter the final EROI result. The model has been calculated as the solar/battery combination in which there were 438 h (5 % of annual hours) below 50 % of battery capacity. The least cost option used 11.1 kW of solar capacity and 63 kWh of battery capacity assuming 50 % depth of discharge (the most common depth of discharge for lead–acid deep-cycle batteries), equivalent to about two days of storage.

Note that a solar capacity of 4.4 kW would generate sufficient *annual* energy, but off-grid systems must be sized according to winter insolation rather than annual average insolation. Using these data, the EROI can be calculated as 1.3 (see Table 5.4).

The lifetime discounted cost of the system is estimated at $80,317 with a LCOE of 47 cents/kWh (assume solar $2,500/kW, batteries $250/kWh with 7.5-year life, 5 kW off-grid charger/inverter $1,000/kW with 15 year life, 3.5 % discount rate).

5.17 Dynamic EROI of Off-Grid System

The dynamic EROI discussed earlier can also be applied to the off-grid system. Using the assumptions above, it is possible to construct a dynamic EROI curve for the projected life of the system. Despite the system using solar modules with a "raw" EROI of 9.4 (including BOS), the need to oversize the panel area to ensure adequate supply during winter, along with batteries, results in the system being in energy debt for most of its life (Fig. 5.5).

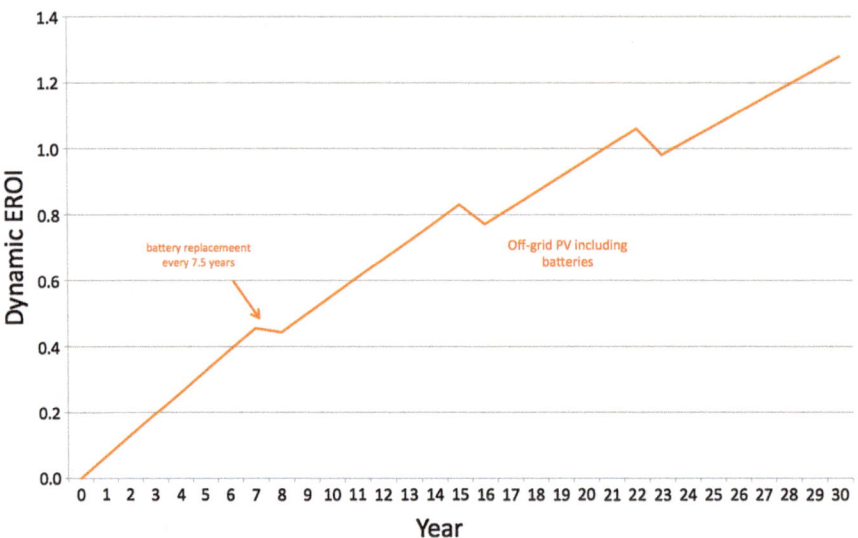

Fig. 5.5 Dynamic EROI of off-grid solar system

5.18 EROI of Concentrated Solar Thermal Systems

There was strong interest in concentrated solar thermal power (CST) in the 1980s and 1990s, mostly in parabolic trough designs, but renewed interest in recent years has explored tower, dish and Fresnel designs. The vast majority of CST plants have been constructed in the USA and Spain, with 93 % of the 2,800 MW global capacity being parabolic trough designs. As of 2013, another 2,500 MW is under construction, with most capacity continuing to be parabolic trough.

The primary strength of solar trough and the similar Fresnel design is supplying peak and intermediate loads during summer in regions with strong sunshine and clear skies (International Energy Agency (IEA) 2010a). But more recent solar tower designs with thermal storage and natural gas backup offer near-baseload operation. In this regard, from an EROI systems perspective, trough and Fresnel are more similar to PV except that the thermal inertia alleviates the rapid ramping and cloud flicker inherent in PV, and trough has the option of using natural gas (or other) backup as a means of improving the economic viability of the plant. In practice, the low efficiency of solar-based gas-fired generation usually precludes its use outside of supporting solar generation.

Electricity generation of PV modules is roughly proportional to the number of photon strikes, irrespective of whether the insolation is direct or diffuse (i.e. filtered or reflected by clouds). On the other hand, CST can only focus direct sunlight and require that the working fluid exceeds a minimum usable temperature before electricity generation can begin. On a summer day, the first hour or so will be taken up with raising the fluid temperature, after which a proportion of all additional heat will be converted into electricity.

However, during winter, a combination of cloud cover and low altitude sun combine to reduce the available insolation. On many days, the working fluid will never exceed the temperature threshold, and even clear winter days will produce only a fraction of the typical summer output (Odeh et al. 1996). This is particularly problematic for solar troughs, which only have single-axis tracking, and in which the working fluid is distributed throughout the entire installation and subject to heat losses related to wind speed, absorber emissivity, absorber temperature, and ambient temperature (Odeh et al. 1996). Due to the much higher concentration factor and two-axis tracking, solar towers perform better during winter.

Dale (2013) provides a lifecycle cost equating to an EROI estimate of around 25 for solar trough based on 13 studies, all of which used process analysis. When backup was included to allow a trough system to operate either 12 or 24 h day, Mujica (2009) calculated the resulting EROI at between 3 and 5.5 in Chile. This aligns with studies of PV, in which backup has a significant impact on EROI.

There have been fewer solar towers constructed, and this is reflected in the dearth of LCAs. Of studies that examined solar tower, Zhang et al. (2012) produced an EROI of 6.9 for the 1.5 MW solar tower in Beijing China. Most cost estimates for solar tower are greater than troughs, in part due to the more mature market status of troughs, but also a reflection of the difference in construction of central receivers due to geometry, mirror precision, and atmospheric attenuation.

From an EROI perspective, the central receivers are more important since troughs are not likely to take on a role beyond peak and intermediate load in sunny regions that have summer peaking grids. If CST were to take on a meaningful role, it will most likely be central tower since these can generate outside of summer and have more scope for storage and higher efficiency.

5.19   Additional Transmission Infrastructure for CSP

Unlike distributed PV, CST systems will generally require additional transmission infrastructure. Given that the ideal climate for CST is arid and inland to avoid cloud and humidity, supply centres will usually be located remote from demand centres, which in Australia, are mostly located near the coast. Taking (optimistic) projected estimates of the cost of CST in 2030 of around AUD $4,500 per kW in 2030 (at 2013 dollars) (see Table 20 Australian Energy Market Operator (AEMO) 2013), with cost of transmission to service CSP plants at an average $1,000 per kW of capacity (Table 25 Australian Energy Market Operator (AEMO) 2013), the additional cost of additional transmission can be estimated at around 20 %.

5.20   Case Study: Materials Requirements for a Large-Scale CST Roll-Out

There have been three recent Australian studies that detail proposals for a 100 % renewables electricity system. All consist mostly of a suite of solar CST, solar PV, wind, biomass, and hydro. CST has taken a central role in all of the plans.

Table 5.5 Recent Australian renewable energy plans using concentrated solar thermal power

Study	Timeframe	CST (GW)
Elliston et al.	Simulation for 2010	15.6
AEMO 100 % renewables study	2030—scenario 1	12.5
Beyond Zero Emissions (BZE)	2020	42.0

Table 5.6 Material flows of CST based on NEEDS life-cycle inventory

	BZE—42.0 GW (tonnes)
Concrete	95,300,000
Steel	33,300,000
Flat glass	9,000,000
Copper	100,000
Storage salts (nitrates)	34,800,000

These are interesting because, if taken at face value, would require massive volumes of materials that are all heavily dependent on the fossil-fuelled industrial enterprise and provide limited scope to retreat from reliance on fossil fuels for their construction and ongoing replacement. Rather than focusing on the embodied energy or the greenhouse intensity of the plants, it is interesting to consider the basic material requirements of the roll-out due to the sheer scale of such proposals (Table 5.5).

In order to assess the material flows of the CST, the authoritative NEEDS-project provides material and energy flow for the Spanish "Solar Tres" 15 MW solar tower project (later renamed Gemasolar 19.9 MW plant) (see page 88 New Energy Externalities Developments for Sustainability (NEEDS) 2008). This has been scaled to the capacity of the BZE plan, which has favoured the use of high capacity factor solar tower, and provides a current baseline.

The use of CST as a baseload source requires a capacity factor of greater than 70 %. This requires a solar multiplier of typically greater than 2.5 and more than 12 h storage (i.e. a 100 MW solar field but a 40 MW power block); these are plants at the high end of materials and costs at greater than USD $9,000 per kW (see Table 4.1 International Renewable Energy Agency (IRENA) 2012). The indicative estimates are intended as a guide to scale and exclude additional transmission infrastructure and the significant costs associated with shipping and construction in remote regions (Table 5.6).

The material flows for concrete, steel, flat glass, copper, and storage salts dominate solar thermal construction. Given that the concentration of sunlight onto a central receiver requires precision tracking, structural rigidity of support structures, and precise mirror dimensions, the use of significant materials in the heliostat structure and foundation is not surprising. Also, given that there are two light paths (incoming from the sun plus reflected beam to the central receiver), the heliostats are usually spaced in a "stretching spiral" configuration to avoid shading and blocking; this requires a trade-off between cosine efficiency, atmospheric efficiency, and land area since the heliostat spacing becomes considerable beyond a modest system nameplate power.

Future reductions in materials and cost would be expected as the technology develops, but as Trainer (2010a), and Nicholson and Lang (2010) note, assumed cost reductions via the learning-curve should be treated with caution at this stage of development. Although there are many projections of future costs, these are really no better than educated guesses until sufficient capacity has been installed at a commercial scale. Notwithstanding these disclaimers, solar tower is perhaps the most promising of the solar technologies to substitute for baseload generation.

The issue here is that the premature retirement of functional energy assets, many with decades of operational life, with new plants requiring massive material flows demonstrate the contradiction of attempting a shift from fossil fuels. For example, if the BZE plan were implemented over the projected 10-year timescale, Australia's annual steel demand could increase by 50 % as a consequence of the CST roll-out, yet the mining of the iron ore, coal, smelting, processing, transport, and manufacturing are heavily reliant on fossil fuels. For example, there is no practical large-scale alternative to metallurgical coal for smelting the 1.5 billion tonnes of iron that is produced globally each year (Smil 2009)—although a partial reversion to charcoal-based steel smelting is considered feasible with a high enough carbon price (Norgate and Langberg 2009), the subsequent competition for the limited "sustainable" forestry biomass would be intense; charcoal production would be competing with biofuels, paper pulp, greater use of sawn timber for construction, and firewood. There remains scope for improved resource utilization, including recycling and the use of natural gas, but neither of these removes the basic requirement for the mass production of virgin steel based on coal.

Given the harsh desert operating environment, the life of the solar field is typically projected at 25–30 years; hence, a plan of continual replacement or refurbishment would be required (see New Energy Externalities Developments for Sustainability (NEEDS) 2008 and International Renewable Energy Agency (IRENA) 2012). Hence, although the deployment of CST would produce a net reduction in emissions relative to unsequestered fossil fuels, it is still highly dependent on the fossil-fuelled enterprise for its construction and maintenance.

5.21   Case Study: High-Penetration Household Solar PV

This case study considers the broader energy costs associated with household PV. These costs relate to a pro-rata allocation due to distribution and transmission costs, the embodied energy of storage, which will become necessary beyond around 5 % of system energy, and the labor costs associated with the Australian PV industry.

The additional costs associated with the altered voltage profile are not included in the calculations because they are difficult to estimate, would only apply as penetration increases, and are discussed further in Chap. 3.

Once PV penetration reaches around 5 %, an increasing proportion of PV output will spilled or stored. For example, if 25 % of the PV output is cycled through storage at 80 % efficiency, the resulting energy loss equates to 5 %. If 4 h of

Table 5.7 Shares of Electricity supply output and employment by ANZSIC group, 2006–2007

	Industry value added (%)	Employment (%)	Net capital expenditure (%)
Generation	35	22	30
Transmission	11	6	18
Distribution	47	62	48
On-selling and market operation	7	11	4
	100	100	100

Source Productivity Commission (2012). Total value added $20.1B (pg. 35 (Australian Bureau of Statistics (ABS) 2012)), employment 52,000 (Department of Resources Energy and Tourism 2012)

storage are provided at 50 % depth of discharge, a raw EROI of panels and BOS of 9.4 reduces to 3.2, assuming that lead–acid batteries are used.

These continue to be the most popular storage device for household solar, and the modern construction can be dated to Camille Fauré's process in 1881 for coating the lead plates, which opened up the industrial scale production of the battery (Pavlov 2011). The longevity of the lead–acid battery provides a reality check on the limits of technological innovation in energy conversion and storage (for example, Eisler (2009) 50-year historical account of the hydrogen fuel cell provides an antidote to the notion that a revolution in storage is "just around the corner").

Since homes with solar PV still require the distribution network for reliable power, along with the normal service and billing overhead of electricity supply, a pro-rata allocation of these costs should be applied if solar PV is to achieve an energy surplus; otherwise, the PV would represent an energy burden on the network. These costs account for around two-thirds of the average Australia electricity bill and are broadly consistent with the national costs of the electricity supply industry (see Table 5.7). Hence taking a simplified cost-based embodied energy content of generation versus remaining costs would imply that the raw EROI of solar PV, with the boundary at the inverter output, must be reduced by half (Fig. 5.6).

Further energy costs that are beyond the conventional boundaries of PV LCAs are the labor costs of the Australian PV industry, including installation, legal, bureaucratic overheads, and other costs. Watt et al. (2011) report the estimated "PV-related labour places" in Australia for 2011 at 10,600. System, installation, manufacture, and distribution make up 7,100, with financial, legal, REC traders, consultants, and analysts making up 3,000 and research and development including 300.

Although not included in these calculations, the regulatory and bureaucratic overhead associated with the promotion and administration of renewable energy and climate change policy represents a significant cost relative to the installed capacity of renewable energy. In 2010–2011, the DCCEE 2012b, for example, had operating expenses of $219 million and 7,000 airplane flights were noted in

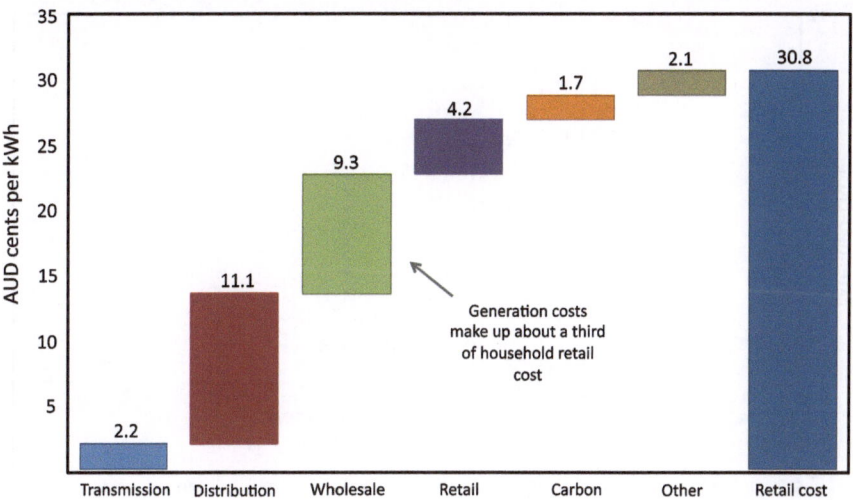

Fig. 5.6 Projected retail residential electricity price components 2013–2014, Australia. *Source* Productivity Commission (Australian Productivity Commission 2013)

the annual report. In addition, every state in the federation, along with many local councils, has their own bureaucratic overhead associated with the promotion of renewable energy and energy efficiency. There are also compliance costs associated with greenhouse accounting for industry.

The role of the Federal DCCEE is interesting because it has an explicit purview to expand the role of renewables and energy efficiency, but has no role in the consideration of the two major global low-emission energy sources, large-scale hydro and nuclear, nor the technology often considered essential in climate change policy, coal with carbon capture (see Department of Climate Change and Energy Efficiency 2012c). This last point is important since the DCCEE is essentially a non-profit extension of the renewable energy industry; techno-renewable energy is assumed *ipso facto* to be the solution to climate change (see Department of Climate Change and Energy Efficiency 2012a and Climate Commission 2013).

Within the context of international concern with climate change, it is appropriate that Australia undertakes climate activities, including research, greenhouse measurement, reporting, and international cooperation and negotiations. But unlike other essential bureaucratic and government functions required to maintain a civil society, it is not obvious that Australia is wealthier, more productive, has a cleaner environment, or benefits in a tangible way from funding these roles. Given that an estimated two-thirds of industry solar PV value was expended on imports in 2011 (Watt et al. 2011), and similarly for wind turbines, the promotion of renewables expands Australia's current account deficit while also increasing energy costs.

It can be argued that the bureaucratic overhead of climate policy in Australia is modest in relation to whole-of-government costs, but the nature of bureaucracy is that expansion is usually easier than contraction; climate policy has created an

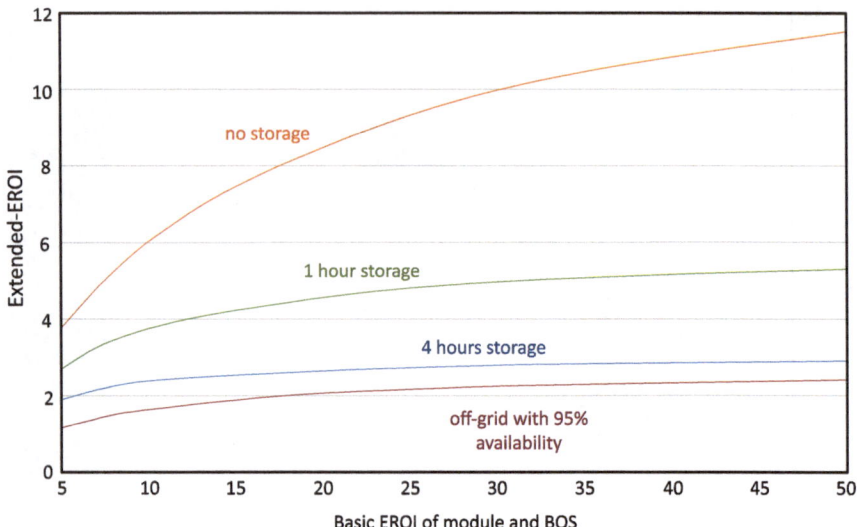

Fig. 5.7 Extended-EROI of PV systems versus basic-EROI. No storage include pro-rata allowance for distribution costs and employment in PV-related industries. *Source* Palmer (2013)

industry of stakeholders and rent seekers, including consultancies, NGOs, and interest groups that directly benefit from the expansion of climate bureaucracy, which in turn contributes to an entrenched groupthink of enhanced regulation, larger government, and greater accountability to international institutions. This aligns with Tainter's (1990) thesis that societies become more complex to solve problems, but in doing so place a greater burden on the productive elements of society.

Although each of these additional energy costs represents individually a small proportion of the energy generated by PV, taken together, they reduce the extended EROI significantly. The key insight is that the additional downstream energy impacts included in the extended EROI overwhelm the embodied energy of the panels. It confirms that the strong focus on developing a superior solar cell over-looks the observation that the extended EROI will always be constrained by the weakest links. The most critical problem is that PV requires storage to increase its value to the network, but storage significantly undermines its energy return, in contrast to conventional electricity generation in which the storage is built into the primary energy source (i.e. coal, gas, uranium, hydro) (Fig. 5.7).

References

Alstone P. Embodied energy and off-grid lighting. Berkeley: Lawrence Berkeley National Laboratory; 2012.

Australian Bureau of Statistics (ABS). 5206.0 Australian national accounts: national income, expenditure and product. Canberra: ABS; 2012.

Australian Energy Market Operator (AEMO). 100 percent renewables study—modelling outcomes. Melbourne, Australia; 2013.

Australian Productivity Commission. Electricity network regulatory frameworks report. Canberra: Productivity Commission; 2013.

Blakers A, Weber K. The energy intensity of photovoltaic systems. Centre Sustain Energy Syst Aust Nat Univ. 2000.

Bossel U. The hydrogen illusion: why electrons are a better energy carrier. Cogener On-site Power Production. 2004; 55–9.

Bundesministerium für Wirtschaft und Technologie. Generating capacity, gross electricity generation and gross consumption Germany. Available online: http://www.bmwi.de/BMWi/Redaktion/Binaer/Energiedaten/energietraeger10-stromerzeugungskapazitaeten-bruttostromerzeugung,property=blob,bereich=bmwi2012,sprache=de,rwb=true.xls (2013). Accessed 1 Jan 2013.

Cleveland CJ. Net energy from the extraction of oil and gas in the United States. Energy. 2005;30(5):769–82.

Cleveland CJ, Costanza R, Hall CAS, Kaufman R. Energy and the U.S. economy: a biphysical perspective. Science. 1984;225:890–7.

Climate commission. The critical decade: Australia's future—solar energy; 2013.

Crawford RH. Validation of a hybrid life-cycle inventory analysis method. J Environ Manage. 2008;88(3):496–506.

Crawford RH, Stephan A. The significance of embodied energy in certified passive houses. World Academy of Science, Engineering and Technology. 2013. p. 78.

Dale M. A Comparative analysis of energy costs of photovoltaic, solar thermal, and wind electricity generation technologies. Appl Sci. 2013;3:325–37.

Deloitte. Advanced metering infrastructure customer impacts study, vol. 1. Melbourne: Department of Primary Industries; 2011.

Denholm P, Margolis RM. Evaluating the limits of solar photovoltaics (PV) in traditional electric power systems. Energ Policy. 2007;35(5):2852–61.

Department of climate change and energy efficiency. Annual report 2011–12. Canberra: ACT; 2012b.

Department of climate change and energy efficiency. Securing a clean energy future: the Australian government's climate change plan. Canberra: ACT; 2012c.

Department of climate change and energy efficiency. 100 percent renewables study—scope. Canberra: ACT; 2012a.

Department of Resources Energy and Tourism. Energy in Australia 2012. Canberra: DRET; 2012.

Eisler MN. A modern philosopher's stone: techno-analogy and the bacon cell. Technol Cult. 2009;50(2):345–65.

Fthenakis V. How long does it take for photovoltaics to produce the energy used. Nat Soc Prof Eng. 2012.

Fthenakis V, Frischknecht R, Raugei M, Kim H, Alsema E, Held M, de Wild-Scholten M. Methodology guidelines on life cycle assessment of photovoltaic electricity. Upton: IEA-PVPS; 2011.

Government Prices Oversight Commission. Bass strait Islands electricity price inquiry. Hobart: Tasmania Government; 2008.

Gupta AK, Hall CAS. A review of the past and current state of EROI data. Sustainability. 2011;3(10):1796–809.

Hydro Tasmania. Currie power station. Available online: http://www.hydro.com.au/system/files/documents/PS_Factsheets/Currie_Power_Station-Fact-Sheets.pdf (2008). Accessed 1 Jan 2013.

Hydro Tasmania. King Island renewable energy project. Available online: http://www.kingislandrenewableenergy.com.au/project-information/diesel-ups (2013). Accessed 1 Jan 2013.

International Energy Agency (IEA). Technology roadmap—concentrating solar power. Paris: IEA; 2010a.

International Energy Agency (IEA). Key world energy statistics—2012. Paris:IEA; 2012.

International Renewable Energy Agency (IRENA). Bonn: concentrating solar power; 2012.
King CW. Net energy principles for understanding if energy production is an economic constraint. Webber Energy Group Symposium; 2013.
Kubiszewski I, Cleveland CJ, Endres PK. Meta-analysis of net energy return for wind power systems. Renew Energy. 2010;35(1):218–25.
Kümmel R. The second law of economics. New York: Springer; 2011.
Lenzen M. Errors in conventional and input-output—based life—cycle inventories. J Ind Ecol. 2000;4(4):127–48.
Lenzen M. A generalized input–output multiplier calculus for Australia. Econ Syst Res. 2001;13(1):65–92.
Lenzen M. Life cycle energy and greenhouse gas emissions of nuclear energy: a review. Energ Convers Manage. 2008;49(8):2178–99.
Ilic, M. Greening the Azores Islands: the key role of dynamic monitoring and decision systems (DYMONDS). Faculty of Technology, Policy and Management Delft University of Technology; 2011.
Mujica L. Net energy analysis of hybrid concentrated solar thermal power plants in Chile: a selection methodology for optimal plant location based on sustainability attributes. 2009.
Murdock D. Wind—(hydro)—diesel power systems: Flores and Graciosa Island, Azores Portugal. PB New Zealand Ltd; 2012.
Murphy DJ, Hall CAS, Dale M, Cleveland C. Order from chaos: a preliminary protocol for determining the EROI of fuels. Sustainability. 2011;3(10):1888–907.
New Energy Externalities Developments for Sustainability (NEEDS). Final report on technical data, costs, and life cycle inventories of solar thermal power plants: project 502687; 2008.
Nicholson M, Lang P. Zero carbon Australia—stationary energy plan critique. Available online: http://bravenewclimate.files.wordpress.com/2010/08/zca2020-critique-v2-1.pdf (2010). Accessed 1 Jan 2013.
Norgate T, Langberg D. Environmental and economic aspects of charcoal use in steelmaking. ISIJ international. 2009;49:587–595.
Odeh SD, Morrison GL, Behnia M. Thermal analysis of parabolic trough solar collectors for electric power generation. In: Proceedings of ANZSES 34th annual conference, Darwin, Australia; 1996. p. 460–467.
Palmer G. Household solar photovoltaics: supplier of marginal abatement, or primary source of low-emission power? Sustainability. 2013;5(4):1406–42.
Pavlov D. Lead-acid batteries: science and technology: science and technology. Oxford: Elsevier Science; 2011.
Prieto PA, Hall CAS. Spain's photovoltaic revolution: the energy return on investment. New York: Springer; 2013.
Productivity commission. Productivity in electricity, gas and water: measurement and interpretation. Canberra: PC; 2012.
Raugei M, Fullana-i-Palmer P, Fthenakis V. The energy return on energy investment (EROI) of photovoltaics: methodology and comparisons with fossil fuel life cycles. Energy Policy. 2012;45:576–82.
Shinnar R. The hydrogen economy, fuel cells, and electric cars. Technol Soc. 2003;25(4):455–76.
Smil V. Prime movers of globalization: the history and impact of diesel engines and gas turbines. Cambridge: MIT Press; 2010.
Smil V. The iron age & coal-based coke; 2009.
Tainter J. The collapse of complex societies. Cambridge: Cambridge University Press; 1990.
Tainter JA, Patzek TW. Drilling down: the gulf oil debacle and our energy dilemma. New York: Springer; 2011.
Trainer T. Comments on zero carbon Australia. Available online: https://socialsciences.arts.unsw.edu.au/tsw/ZCAcrit.html (2010a). Accessed 1 Jan 2013.
Treloar GJ. A comprehensive embodied energy analysis framework. 1998.

Watt M, Passey R, Johnston W. PV in Australia 2011: prepared for the IEA cooperative programme on PV power systems. Liberty Grove: Australian PV Association; 2011.

Weißbach D, Ruprecht G, Huke A, Czerski K, Gottlieb S, Hussein A. Energy intensities, energy returned on invested (EROIs), and energy payback times of electricity generating power plants. Energy. 2013.

Wenham SR, Green MA, Watt ME, Corkish R. Applied photovoltaics. Sydney: UNSW Centre for Photovoltaic Engineering; 2006.

Zhang M, Wang Z, Xu C, Jiang H. Embodied energy and emergy analyses of a concentrating solar power (CSP) system. Energy Policy. 2012;42:232–8.

Chapter 6
Driving Down Emissions: The Role of Carbon Pricing

6.1 Carbon Pricing as the Dominant Policy Tool to Reduce Emissions

Much of the discussion of climate change rests on the assumption that excess carbon dioxide emissions are a consequence of a market failure. Since the long-run costs of climate change are not built into the price of fossil fuels, the assumed solution is to internalize the costs of carbon. Indeed within the international policy community, carbon pricing is the favoured policy of neoclassical economists (e.g. Garnaut, Stern) and dominates policy discussions (Geels 2012).

The concept of a price on greenhouse emissions has a compelling logic; all other things being equal, a price on carbon will reduce carbon relative to other economic indicators. The European experience with carbon taxes since the 1990s has been that carbon taxes do in fact reduce emissions from what they would have otherwise been, and mostly without a significant loss of industrial competiveness (Andersen 2010). Indeed, even putting aside the issue of climate change, there is an argument that shifting the tax burden from income taxes in favour of consumption taxes will tend to encourage savings and investment (Mankiw 2009).

But unlike the pricing of social ills or pollutants, the purpose here is not just to marginally reduce the quantity of emissions, but change, in a fundamental way, the energy systems that *enable* the advanced societies, along with the health, education, industry, and culture that is modernity.

There is little argument against renewable energy providing an abatement role within a fossil-fuelled economy; indeed, carbon pricing is likely to be an effective marginal *abatement* strategy in the advanced economies. But it is less clear that it can be an effective global *energy* strategy in an era of lowering energy return on investment and the need to bring affordable energy to much of world's population. A discussion of the role of carbon pricing provides an opportunity to draw out some of the challenges of shifting to a high-penetration renewables scenario within a global context.

G. Palmer, *Energy in Australia*, Energy Analysis,
DOI: 10.1007/978-3-319-02940-5_6, © Graham Palmer 2014

6.2   Carbon Pricing as a Proxy for the "Soft Energy Path"

Although carbon pricing is advocated as a policy instrument to drive abatement through both the demand effect (i.e. efficiency and conservation) and the substitution effect (i.e. low-emission energy and fuel switching), more often it embodies a particular world view that supports a variant of the "soft energy path", originally advocated by Lovins (1976). This includes much stronger regulations for technical efficiency, a shift towards localized and distributed generation, mandatory renewable energy targets, and a stronger role for international institutions.

For example, consider Sweden and France, which have among the lowest-cost electricity in Europe with a near-zero emission intensity for electricity supply of 23 and 71 g CO_2/kWh, respectively, compared to for example, Germany at 672 g and Denmark at 375 g CO_2/kWh (Brander et al. 2011). Yet Sweden and France are still subject to the European Commission requirements for expanding renewable energy and emission targets (European Commission 2013).

Similarly, in Australia, despite the Department of Climate Change and Energy Efficiency being the peak national body for climate change, as noted earlier, it has a specific purview to support specific techno-renewable energy sources, such as wind, solar, tidal, and geothermal.

6.3   Carbon Pricing Standard Model

The standard model for carbon pricing is a low starting price, with a steadily rising price over time. It is generally taken *ipso facto* that the ratcheting upwards of the carbon price will propel a commensurate ratcheting downwards of lower-emission energy sources in a virtual dance until energy supply reaches low or near-zero emission.

The critical question for neoclassical economics becomes determining the optimal pathway to achieve the optimal rate of decarbonization within the context of climate uncertainty, without imposing unnecessary burdens on economic development (Nordhaus 2007). It is also assumed that, by default, carbon abatement can be imposed by economic contraction (i.e. a recession) in the absence of low-emission options.

The strength of an explicit carbon price versus alternative policy instruments is that it targets emissions directly while remaining technology neutral; it is taken as given that holding all other things equal, a price on CO_2 will reduce the quantity of CO_2 relative to other economic indicators while avoiding rebound factors (Alcott 2009). Since technology-specific policy instruments are rarely the most effective abatement tools, an explicit carbon price would be expected to achieve greater abatement at lower cost.

In a comprehensive review of Australian greenhouse policies, Wood and Edis (Wood and Edis 2011) found that market mechanisms have been far more effective

Fig. 6.1 PV systems, energy versus abatement

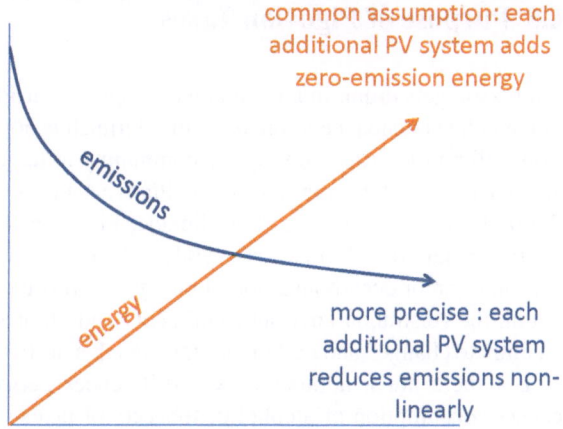

than subsidies and other policy instruments at reducing emissions. Similarly, the Australian Productivity Commission (Australian Productivity Commission 2011) noted that an explicit carbon price would be cost effective at a low carbon price since it captures a considerable amount of low-cost abatement on the demand side (i.e. energy efficiency and conservation). Australia has relatively high per-capita energy consumption, so it is not surprising that there is significant scope for low-cost abatement given appropriate incentives.

6.4 Marginal Abatement Versus Deep Abatement

The underlying assumption is that there is a more-a-less linear relationship between marginal abatement activities and actual abatement (i.e. adding a thousand solar panels or wind turbines to an electricity system results in a thousand times more abatement than adding a single panel) and that this linearity can be extrapolated *ad infinitum* until some technical boundary is eventually reached.

Indeed, the early experience of the expansion of wind and solar is that at low penetration, a linear approximation provides a reasonable first estimate. The problem is that wind and solar PV force increasing demands on ancillary services and contribute to adverse interactions with conventional thermal plant; hence, the linearity assumption breaks down beyond a given penetration of intermittent plant, typically from around 5 % of annual system penetration (Inhaber 2011 and Palmer 2013). Put another way, it is often assumed that each additional solar system cumulatively adds zero-emission electricity to the network, but a more accurate explanation is that each solar system reduces emissions slightly from what they would otherwise have been in a nonlinear fashion (Fig. 6.1).

6.5 Purpose of Pigovian Taxes

If a person gets drunk in a bar and then staggers outside and gets in a fight, the cost of the police attendance to break up the skirmish is not borne by the bar that profited from selling the drinks, but by the community at large; hence, the rationale behind alcohol taxes, which are a form of Pigovian tax after Arthur C. Pigou in 1920. Similarly, in the case of carbon dioxide, Pigovian taxes are intended to internalize the social cost of emissions, while reducing carbon emissions through reduced consumption or decarbonization of energy. Such is the appeal of carbon pricing that within the Australian environmental NGO and climate activist community, "action on climate change" and carbon pricing are taken as two sides of the same coin.

But there are a number of key differences between applying a "sin tax" to excess consumption of alcohol or tobacco, or taxes on conventional pollutants. In most cases, the technology or strategies already exist to reduce the effect of the offending substances to manageable levels; internalizing the cost provides a policy instrument to enable the application of those abatement technologies. Nextly, in most cases, the technology or accounting system is available to readily measure the offending substance or activity. And most problems addressed through Pigovian taxes are usually related to "tame" problems involving specific products or activities with specific outcomes.

But the carbon dioxide problem is quite different to conventional problems addressed with Pigovian taxes. Few problems associated with conventional pollutants or social ills have an explicit objective of nearly completely nullifying the problem in the long run at the global level, without there being a socially acceptable substitute readily available. Few pollutants are as pervasive and ubiquitous as carbon dioxide, and CO_2 is not a pollutant in the conventional sense. Carbon dioxide and the other two principle greenhouse gases (excluding water vapour), methane and nitrous oxide, are naturally occurring trace gases; the issue is the carbon cycle's sensitivity to anthropogenic perturbations (Smil 2007). Emissions are associated with energy, agriculture, land use, microbes, and forestry, and although many energy sources can be accurately measured or estimated, there is no practical means to reliably measure the multitude of point sources at a local level. The case of Australia's extensive land clearing through the 1990s provides ample evidence of the challenges of controlling land clearing and the associated emissions, despite advanced satellite monitoring technologies and an advanced legislative and bureaucratic oversight (Macintosh 2012).

Nextly, it is not clear what the most appropriate use of the taxes should be. In Europe, €25 billion annually has been shifted from labour-related taxes to carbon–energy taxes (Andersen 2010). On the other hand, some advocates recommend the use of carbon taxes to fund renewable energy schemes. In Australia, the broad-based carbon tax of AUD $23 per tonne CO_2-e during 2013 was recycled as rebates for householders and support for energy-intensive industry.

Few proposed schemes recommend the conventional role of Pigovian taxes, that being to fund the cost of externalities, or in the case of carbon dioxide, funding the cost of climate adaption; to date, there is little evidence that climate change has increased the normalized economic loss from natural disasters (Neumayer and Barthel 2011; Simmons et al. 2013).

6.6 The US Sulphur Dioxide Scheme as a Model for Emission Trading Schemes

The Kyoto protocol endorsed the emission trading (ETS) mechanism as the preferred carbon pricing model, since it was said to provide the most political feasible model within a multi-lateral context. The ETS was directly modelled after the successful US sulphur dioxide scheme, in part due to the effective lobbying of the US Clinton Administration during the Kyoto negotiations, and the acknowledgement that harmonized national carbon taxes were not politically feasible (MacKenzie 2009). A comparison of the sulphur dioxide scheme clarifies some of the key differences between the simple price mechanism of sulphur dioxide and any comparable scheme to reduce greenhouse emissions.

The concept of emissions trading for controlling pollution was proposed in the 1960s by Crocker and Dales, but the first formal analysis was given by Montgomery in 1972 (Hanemann 2009). The emergence of pollution trading emerged in an environment of neoliberalism, which included economic liberalization, privatization, and deregulation; indeed, the concept of trading pollution rights fitted neatly into the neoliberal paradigm (Lohmann et al. 2006). This very feature of trading schemes, and the resulting alliance between the notionally right-wing financial industry and notionally left-wing climate campaigners and environmental NGOs, was one of the defining features of post-Kyoto support for trading schemes (MacKenzie 2009).

The sulphur dioxide program of the Clean Air Act of 1990 was a response to the problems of acid rain and urban smog. But the SO_2 scheme was different to proposed carbon dioxide schemes in some fundamentals ways. The SO_2 scheme targeted a specific pollutant that could be directly measured at a limited number of point sources, and the technology to limit emissions was already known. Further, the program was technology neutral and operated within the established framework of the electricity supply industry relying almost exclusively on synchronous generators driven by one of three basic types of turbine (hydro, Rankine, and Brayton).

Under a command-and-control approach, coal-fired generators would have been required to install sulphur scrubbers to ensure every generator complied with an emission standard (there are also prescriptive unit limitations from 1970 to prevent local adverse health outcomes).

But the benefit of the pricing scheme was that the market was able to self-select the most cost-effective method, including greater use of nuclear and gas, changing the dispatch order, and the use of lower sulphur coal. Importantly, a measurable reduction in SO_2 at the flue resulted in unequivocal abatement. The trading aspect allowed generators with low SO_2 generation to sell excess permits and therefore maximize sulphur reductions from the lowest-cost sources. But despite the broad options available, Ellerman and Dubroeucq (2004) estimated that about 85 % of the reduction in sulphur emissions between 1994 and 2002 was associated with a reduction in emissions at individual generating units (i.e. cleaning up old plants).

6.7 Renewable Energy Targets and Carbon Pricing may Conflict

Nearly, all CO_2 schemes operate in parallel with other schemes, especially renewable energy targets. Although it is often assumed that the long-run effect of both policy instruments will be to encourage renewable energy, the policy instruments are targeting different metrics and distort the normal operation of the electricity market. For example, in Australia, PV operates outside of the normal dispatch queue without any direct control from system operators. Wind power operates within the normal market, but is operated on a must-take basis, regardless of prevailing market conditions.

In most cases, solar energy displaces lower-emission gas-fired generation rather than higher-emission coal-fired generation since coal-fired baseload resides at the bottom of the dispatch queue. Since solar PV only generates during the day, there will always be peak or intermediate load supply higher in the dispatch queue; hence, it mostly displaces generation that is already contributing to lowering the system emission intensity. In contrast, wind energy may also displace coal due to the prevalence of night-time wind (when the entire load is met with baseload), but thermal cycling may erode some of the emission gains of wind. In many scenarios, it is clear that policy prescriptions can seek to maximize renewable penetration or maximize abatement, but not both.

6.8 Carbon Price Targets a Single Metric but Effects are Multi-Faceted

Nearly all global electricity is generated with synchronous generators driven by one of three basic types of turbine; Rankine-cycle steam turbines fuelled by coal, gas or nuclear, Brayton cycle gas turbines, or water-powered hydro. The same synchronous plant that traditionally supplied energy also supplied the suite of ancillary, security, and stability functions, including primary frequency control, inertia, reactive power support, voltage regulation, capacity, and other essential roles. In the main, these functions have been traditionally provided at low or no cost by conventional synchronous generators.

The price–technology ratcheting postulate that is implicit in carbon pricing theory assumes high substitutability and equivalence of energy sources. But solar

PV, wind, and other renewable energy are intermittent, non-synchronous generators (concentrated solar thermal and biomass-fired boilers being notable exceptions) and cannot meaningfully contribute to the essential security and stability functions without built-in storage. Since carbon pricing only targets system energy, rather than capacity, inertia, other functions, alternative markets, market rules, or other services will need to be developed to account for a shift in mix towards non-synchronous generation. This will necessarily add complexity, oversight, and costs that lie outside the scope of carbon pricing. For example, a number of jurisdictions have explored the necessity of capacity markets, to essentially pay conventional generators to remain online, in order to ensure system security in response to increasing penetration of intermittent generation [see for example UK Department of Energy and Climate Change (UK Department of Energy and Climate Change 2012)].

6.9   Embodied Energy Costs of PV

The energy costs of constructing a low-emission energy system are assumed endogenous to the system and relatively small (i.e. a high EROI). In theory, with a unified and broad-based global carbon price, the embodied energy content of renewable technologies should be incorporated into the energy cost. However, nearly all of the critical components of Australian PV systems are imported; hence, the energy cost of systems is not fully accounted for in a carbon price. Indeed, with few exceptions, the embodied greenhouse emission content of renewable generation is not accounted for when ascertaining their relative competitiveness in the context of a carbon price.

In Australia, only scope 1 (greenhouse releases as a direct result of an activity) and scope 2 (greenhouse emissions related to the consumption of electricity, heat, or steam) are counted in greenhouse reporting (Department of Industry Innovation Climate Change Science Research and Tertiary Education 2013). In the case of fossil fuel energy, these provide a simple and practical methodology for greenhouse compliance, but they exclude the emissions associated with the manufacture of renewable energy systems, which constitute most of the emissions associated with renewable energy.

6.10   Carbon Price as an Innovation Driver

The interaction between the wholesale market and carbon pricing is such that only generation at the margins is affected; hence, low emission generation does not become a "little more competitive" with each ratchet upwards of the carbon price; it is only as the carbon price approaches the margins that operational decisions are affected. Further, the carbon cost applies to every unit of energy whether or not that energy source is close to the margins.

It is often assumed that carbon price will drive innovation for low-emission generation through the "announcement effect" of high future carbon prices. For example, Australian Treasury (Australian Government 2012) modelling assumes a carbon price between AUD \$29 and \$62 per tonne CO_2-e in 2020, rising to \$131 to \$275 per tonne CO_2-e in 2050 (AUD 2010). But since carbon pricing is determined by government fiat, the credibility of commitment creates risks for low-emission innovations (McKibbin and Wilcoxen 2007).

Given that European allowances are trading at less than five € per tonne (USD 6.00) in 2013 and none of the five largest emitters (US, China, Russia, India, Japan) have implemented carbon pricing beyond limited schemes with modest coverage and prices, the medium-term outlook for the major emitters remains weak. Further, since electricity is an undifferentiated commodity product, there is no clear reward for first-movers (Wood and Edis 2011). Even in the event of a strong Australian bipartisan commitment to abatement, future governments may be motivated to subsequently lower the carbon price if competitive low-emission baseload technologies emerge. Since innovators are aware of this, the "announcement affect" of high future carbon prices will tend to be discounted (Ergas 2012 and Montgomery and Smith 2007).

6.11 Lessons from a Leviathan Tax

But there is a further problem with a rising carbon price. Tol (2012) defined the "Leviathan tax" as the hypothetical short-run maximum carbon tax that is budget-neutral (i.e. all other taxes are reduced to zero and replaced by a carbon tax). The Leviathan tax is calculated from the CO_2 emissions, CO_2 intensity of the economy, and the total tax revenue for the country, excluding taxes that directly finance social security programs.

For developed countries such as Australia, the United Kingdom, and the United States, the short-run theoretical Leviathan tax can be above \$200 per tonne CO_2 since the advanced nations collect substantial taxes and also generate substantial economic activity per unit of energy, even if the energy is based on high-emission fuels. Hence, the Leviathan tax does not pose a fundamental constraint on the short-run carbon price for these countries.

On the other hand, a \$1-per-tonne carbon could fund the entire government budgets of Nigeria and Liberia. In the case of the first, third, and fourth highest emitting countries, China, Russia, and India, the Leviathan tax has been calculated by Tol using 2005 data as \$29, \$36, and \$45, respectively (USD 2000). (Table 6.1).

Even in a scenario where these countries were to commit to an aggressive carbon price, the need to ensure a diverse taxation base precludes the possibility of a carbon tax making up more than a minor share of total tax revenues. Hence, the Leviathan tax suggests that the advanced economies are capable of sustaining a much higher carbon price for a given national emission coverage.

Table 6.1   Leviathan tax for the five largest emitters, along with Australia and United Kingdom. Data are constant 2000 US dollars. Source: Tol (2012)

	Leviathan tax $/tonne CO_2	Tax revenue (% GDP)	Carbon intensity kg CO_2/$	CO_2 emissions M tonne CO_2/yr
China	29	8.7	2.76	5,790
United States	223	11.2	0.48	5,595
Russia	36	16.6	4.06	1,616
India	45	9.9	2.08	1,411
Japan	450	10.9	0.24	1,238
Australia	330	24.7	0.73	367
United Kingdom	855	27.2	0.29	542

This raises the problem of participation and the need for other nations opting to undertake proportionally greater effort. Assuming that 50 % of global emissions could be included in a globally unified scheme, Nordhaus (2008) calculated a cost penalty of around 300 %, lowering to around 68 % if 75 % of emissions are included. This reinforces the conclusion that achieving a given climate objective requires a high level of global participation, since much of the potential gains by participating countries will be lost through the non-participants.

6.12   Lessons from the Politics of Carbon Pricing of Australia

Since 2007, five prime ministers and opposition leaders have been defeated or deposed, in large part, due to an inability to construct a coherent narrative on carbon pricing; Conservative Prime Minister John Howard embraced carbon pricing only reluctantly in 2007 and was defeated in the national election to the more progressive Kevin Rudd.

The conservative opposition leader Malcolm Turnbull was then deposed by his own party in 2009 for adopting a too-progressive position on carbon pricing.

Labor Party Prime Minister Kevin Rudd lost public support and credibility when he backed away from carbon pricing following the failure of Copenhagen and was subsequently deposed as prime minister by his own party.

Labor Party Prime Minister Julia Gillard promised that there would be no carbon tax prior to the 2010 national election, but as a response to a hung parliament, formed an alliance with the Australian Greens and legislated a carbon tax, subsequently leading to a dramatic loss of public support and the eventual reinstatement of Kevin Rudd.

Finally, in 2013, the conservative party led by Tony Abbott won the national election, with the rescindment of the carbon tax promoted as a key election pledge.

The problem for the Labor party, in particular, has been that it has been unable to construct a coherent narrative around carbon pricing and has been drawn into the fruitless debates of "targets and timetables". The lessons from Australia are interesting because it demonstrates a broad public willingness to participate in international endeavours to reduce the carbon intensity of the global economy, but an acknowledgement that unilateral action in the absence of a political mandate is unacceptable. Given the current international disposition, it could be argued that a modest revenue neutral scheme that builds an efficient institutional framework, implemented as part of a broader tax reform agenda and climate policy rationalization, would provide a marketable no-regrets policy response.

6.13 Improving Human Welfare will Require More Energy

The human development index (HDI) is an indicator developed by the UN Development Program. It provides a single aggregate statistical index based on life expectancy at birth, education, and gross national income per capita (United Nations Development Program). As such, it provides a useful indicator to the early stages of a nation's development, but does not provide an indication at later stages of development, especially advanced education, healthcare, defence, and science.

There is a strong positive correlation between per-capita national energy consumption and HDI up to around 40 GJ per capita per annum, after which the correlation gradually tapers off. Of the countries with a HDI above 0.9, Ireland and Switzerland have the lowest per-capita energy consumption of around 140 GJ, with the majority of countries with a HDI above 0.8 consuming greater than 100 GJ per-capita per annum. Of the most populous Asian countries, the per-capita energy for China is 76 GJ, India 24 GJ, Indonesia 36 GJ, Pakistan 24 GJ, Vietnam 29 GJ, and Philippines 18 GJ.

Since energy is pervasive in all sectors of the economy, caution needs to be exercised in comparing energy associated with the consumption of final products and the energy consumed in mining and manufacture. For example, Australia is a major exporter of energy and minerals, both of which are energy intensive, but on the other hand, Australia has lagged in energy efficiency indicators (Harrington 2012). Similarly, China is the world's primary manufacturer; an estimated third of China's greenhouse emission are attributed to exports (Weber et al. 2008); hence, a substantial proportion of China's indigenous energy consumption is embodied in manufactured goods for export.

Nonetheless, it becomes readily-apparent that improving human welfare will require a far greater demand on energy in Asia given the large populations of the developing countries, notwithstanding the potential to lower consumption in the advanced countries. A few solar panels supplied by foreign aid, powering a water pump or a radio, can make a great difference to poor villagers in East Timor. But this tells us little about how to build the roads and bridges, the schools and hospitals, or develop a complex civil society (Fig. 6.2).

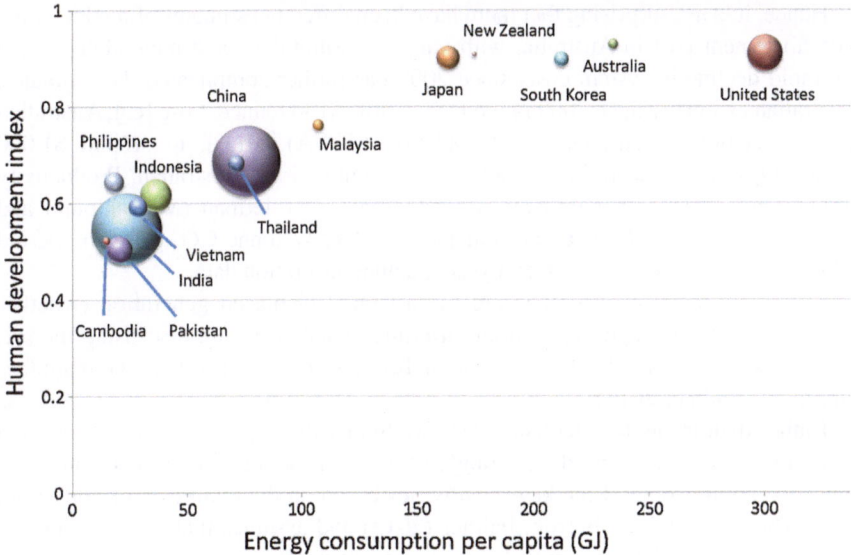

Fig. 6.2 Human development index of selected Asian and other countries versus per-capita energy consumption. Bubble size is proportional to population

6.14 Case Study: Estimating the Abatement Cost of Solar PV

It has been surprisingly difficult to ascertain an abatement cost for solar PV in Australia. The key problem is that abatement cannot be directly measured; we are attempting to estimate notional abatement on the margins relative to our best guess of the counterfactual, extrapolated for the assumed life of the system. In part, the measurement problem is a reflection of the fact that PV has a limited role in substituting for fossil fuel plant; we are simply trying to "green" the fossil fuel grid by reducing the emissions slightly from what they would otherwise have been.

The only practical way to assess the abatement is through estimation or simulation modelling, both of which introduce uncertainty due to the need to employ assumptions and aggregate data across temporal and spatial scales.

For example, for greenhouse accounting purposes, every large Australian generator is assigned a fixed emission intensity, which is calculated from the measured fuel use and electricity sent out on an *annual* basis. But the actual emission intensity of fossil fuel generation in real time is dynamic, and it is difficult to capture the real-time behaviour of generators when they are cycling away from their optimum heat rate. In theory, it could be possible to measure actual coal consumption in real time, but this information is not available, and there has never been a need for such precise granularity; what matters for a power station are the cumulative emissions over a year rather than the potential role of renewable energy in reducing emissions on a second-by-second basis.

Hence, it is not surprising that there have been different estimates of the PV emission abatement cost in Australia, with varying assumptions and methodology, and the rapid decline in system costs since 2009 has further complicated the estimates. Australian estimates since 2009 range from below $100 tonne/$CO_2$-e [e.g. Australian PV Association (Australian PV Association (APVA) 2011)], to around $1,000 tonne/$CO_2$-e [e.g. Australian Productivity Commission (Australian Productivity Commission 2011)]. In Germany, Marcantonini and Ellerman (Marcantonini and Ellerman 2013) calculated an abatement cost of €537 tonne CO_2-e for the period 2006–2010 based on an *ex-post* analysis of actual generation data.

The main areas of difference are the assumed displaced generation (whether coal, gas, or the average), appropriate lifetime, whether we are discussing the private or social cost or subsidies, and the inclusion of emissions due to the manufacture of PV (Palmer 2013).

Rather than trying to establish a decisive figure, the approach of this discussion is to construct a chart based on a range of solar insolations, solar costs, and lifetimes to capture some of the key sensitivities based on the retail cost of household solar. The methodology is from Palmer (2013) and assumes that solar essentially "buys abatement" but does not displace conventional generation capacity, therefore displacing marginal costs [i.e. reduces fuel use and variable operation and maintenance costs (VOM)]. It is assumed that solar displaces gas-fired generation at 600 g CO_2-e/kWh.

6.15 Incorporating the Projected Lifetime into Abatement Cost

Most abatement cost analyses assume a PV lifetime of 30 years, which essentially "deem" the entire 30 years of generation, but it is not clear whether this is the most appropriate methodology. For example, as Fig. 5.3 shows, the abatement should probably be considered as a dynamic rather than a static function. This suggests that some sort of discount function should be built into the methodology in much the same way as cash flows are discounted in financial analysis. Over a period of 30 years, for example, a modest 5 % discount rate applied to the annual abatement would essentially double the cost compared to zero discount.

Further, the longevity is also a serious issue. If a system fails, will householders replace inverters and repair damage out of warranty, such as glazing and bypass diodes? Some of these repairs could easily amount to a couple of year's worth of feed-in tariff. When the home is sold, will the new owners maintain the system—the annual change over rate of Australian homes is 6 % with many homes changing owners in under 10 years.

There are sufficient systems installed from the 1970s to demonstrate a potentially long life, but it is less clear that the thousands of more recent systems, many using "third-tier" Chinese panels will have the same long-run performance and longevity and that householders will have the same passion to maintain the systems.

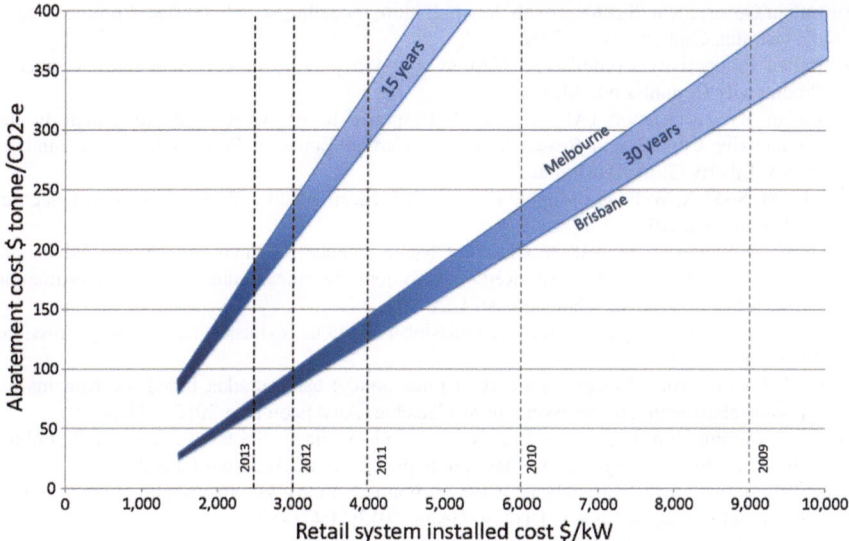

Fig. 6.3 CO$_2$ abatement cost versus system cost for solar PV. *Source* Palmer (2013)

Other issues include dust and grime build-up, shadowing from growing trees, adjacent building developments, and poor siting of panels, etc. These can only depreciate the assumed generation.

Given that only a small number of Australian systems were connected with "gross metering"—in New South Wales and Canberra with some continuing under legacy arrangements—there is no easy way to measure actual solar generation, other than with a field study. Gross metering uses two circuits, but the more common "net metering" uses only a single circuit; hence, the information is lost at the household meter—it would require visually sitting the inverter display to note actual solar generation at regular intervals. Most of the government and PV association estimates of solar generation have been based on estimates rather than actual metered generation and assume more-or-less ideal generation behaviour.

These are all interesting questions that need resolution. Figure 6.3 provides estimates based on deemed 15 and 30 years lifetimes and should be taken as lower bounds since they exclude most of the system-wide costs discussed earlier, discounting, and other factors that will diminish the output.

References

Andersen SM. Europe's experience with carbon-energy taxation. SAPI EN S Surv Perspect Integrating Environ Soc. 2010;3.2.
Alcott B. Impact caps: why population, affluence and technology strategies should be abandoned. J Cleaner Prod. 2009;18:552–60.

Australian Government. Strong growth, low pollution: modelling a carbon price. Commonwealth of Australia, Canberra, ACT; 2012.

Australian Productivity Commission. Carbon emission policies in key economies. Canberra: Productivity Commission; 2011.

Australian PV Association (APVA). APVA Response to PV Costs and Abatement in the Productivity Commission Research Report: Carbon Emission Policies in Key Countries. APVA, Liberty Grove, Australia; 2011.

Brander M, Sood A, Wylie C, Haughton A, Lovell J. Electr-specif emission factors grid electricity. Ecometrica. 2011.

Department of Industry Innovation Climate Change Science Research and Tertiary Education. National Greenhouse and Energy Reporting System: Technical Guidelines for the estimation by faciliities in Australia; Canberra, ACT, 2013.

Ellerman AD, Dubroeucq F. Sources of emission reductions: evidence for US SO_2 emissions 1985–2002. 2004.

Ergas H. Policy forum: Designing a carbon price policy: using market based mechanisms for emission abatement: are the assumptions plausible? Aust Econ Rev. 2012;45(1):86–95.

European Commission. Renewable Energy: action plans and forecasts. 2013. Available online: http://ec.europa.eu/energy/renewables/action_plan_en.html. Accessed 1 Jan 2013.

Geels FW. A socio-technical analysis of low-carbon transitions: introducing the multi-level perspective into transport studies. J Transp Geogr. 2012;24:471–82.

Hanemann M. The role of emissions trading in domestic climate policy. Energy J. 2009;30(2):73–108.

Harrington P. ABC Background Briefing; Energy efficiency: Not in Australia mate! ABC Radio National. 2012. Available online: http://www.abc.net.au/radionational/programs/backgroundb riefing/2012-04-08/3930648. Accessed 1 Jan 2013.

Inhaber H. Why wind power does not deliver the expected emissions reductions. Renew Sustain Energy Rev. 2011;15(6):2557–62.

Lohmann L, Hällström N, Österbergh R, Nordberg O. Carbon trading: a critical conversation on climate change, privatisation and power. Uppsala: Dag Hammarskjöld Foundation, 2006.

Lovins A. Energy strategy: the road not taken? Foreign Aff. 1976; 55:5–15.

Macintosh A. The Australia clause and REDD: a cautionary tale. Climatic Change. 2012;112(2):169–88.

MacKenzie D. Rear Vision: Carbon emissions trading: a way forward or the Emperor's new clothes? ABC Radio National. 2009. Available online: http://www.abc.net.au/radionational/programs/rearvision/carbon-emissions-trading-a-way-forward-or-the/3168000. Accessed 1 Jan 2013.

Mankiw NG. Smart taxes: an open invitation to join the pigou club. East Econ J. 2009;35(1):14–23.

Marcantonini C. Ellerman AD. The cost of abating CO_2 emissions by renewable energy incentives in Germany. 2013.

McKibbin W, Wilcoxen P. Managing price and why a hybrid policy is better for Australia. Committee for Economic Development of Australia, Melbourne, Australia; 2007.

Montgomery WD, Smith AE. Price, quantity, and technology strategies for climate change policy. Human Induced Climate Change: An Interdisciplinary Assessment; 2007.

Neumayer E, Barthel F. Normalizing economic loss from natural disasters: a global analysis. Glob Environ Change. 2011;21(1):13–24.

Nordhaus WD. A question of balance: Weighing the options on global warming policies. New Haven: Yale University Press; 2008.

Nordhaus WD. A review of the "Stern review on the economics of climate change". J Econ Lit. 2007;686–702.

Palmer G. Household solar photovoltaics: supplier of marginal abatement, or primary source of low-emission power? Sustainability. 2013;5(4):1406–42.

Simmons KM, Sutter D, Pielke R. Normalized tornado damage in the United States: 1950–2011. Environ Hazards. 2013;12(2):132–47.

Smil V. Global material cycles. Encyclopedia of Earth; 2007.

Tol RSJ. Leviathan carbon taxes in the short run. Climatic Change. 2012;114(2):409–15.

UK Department of Energy & Climate Change. Electricity market reform: capacity market design and implementation update, annex C; 2012.

United Nations Development Program. Human Development Index (HDI). Available online: http://hdr.undp.org/en/statistics/hdi/. Accessed 1 Jan 2013.

Weber CL, Peters GP, Guan D, Hubacek K. The contribution of Chinese exports to climate change. Energ Policy. 2008;36(9):3572–7.

Wood T, Edis T. No easy choices: which way for Australia's energy future?. Melbourne: Grattan Institute; 2011.

Conclusion

There are a number of lessons to be drawn from a detailed assessment of solar. The first is that context matters. A solar panel installed on King Island that reduces diesel consumption is going to have a different EROI to the same solar panel that is reducing emissions from a coal-fired power station. This may seem odd, but in both cases the solar is providing a supportive or supplementary role, rather than a primary role. In neither case is the panel actually replacing anything. In most cases, the EROI is much lower than commonly assumed, and will usually lie below the critical minimum EROI.

The related lesson is that the net-energy from solar can only be assessed with a whole-of-system approach. It is meaningless to consider the EROI as a stand-alone component since they are almost never used in this way. Nearly every solar panel in the world is used either within a grid-connected system—either as a solar farm or on the roof of properties, in which case the property is still completely dependent upon the grid for more than 60 % of annual hours. Or alternatively, the panel is part of an off-grid stand-alone system with oversized solar capacity (to ensure adequate winter operation) and batteries, and often with a backup gasoline or diesel generator. Other than very minor niche roles, such as powering a water pump in the developing world, solar is always part of a larger system.

Although it is often assumed that solar can be incorporated into a "suite of renewables" with smart-grids to achieve some sort of optimized synergy, the reality is that this imagined synergy rarely exists. There are certainly always going be correlations that improve the usefulness of solar energy—solar powered cooling for example—but at a system-wide scale, these correlations represent the exception rather than the norm. A really smart grid could certainly maximize the usefulness from renewables, but winter performance and the "big gaps" problem will always be a limiting factor (Trainer 2012).

One could easily argue that the EROI of fossil fuels should also reflect the fact that they are also part of a larger system. This is a valid issue and perhaps adjustments would be appropriate, but there is an asymmetric dependence between solar and fossil fuels (Prieto and Hall 2013)—all of the solar systems could be switched

G. Palmer, *Energy in Australia*, Energy Analysis,
DOI: 10.1007/978-3-319-02940-5, © Graham Palmer 2014

off for a day and few would notice. Yet in our modern world, even minor disruptions to communications, the internet, rail stoppages, gasoline supplies, and others can have a significant impact on daily living.

This is not an argument for abandoning solar power, but rather an argument for a more nuanced and targeted role. For example, one of the more useful roles for solar would be to provide network support for summer-peaking electricity grids, provided around 4 h storage is built into the system.

In the future, the development of an ultra-cheap storage device that can be cheaply paired with solar that is embedded in the distribution network would almost certainly represent a disruptive innovation. Concentrated solar thermal with built-in storage is promising—but cost, scale, and the perennial solar problem of poor winter performance are limiting factors.

This leads to the question—what's next?

The declining energy surplus of fossil fuels will be a defining feature of energy systems in the twenty-first century. This will almost certainly have consequences for the dominant liberal economic model we live under, and the assumptions of debt, growth, and cheap energy. A better understanding of the relationship between energy, ecology and society is going to be essential if we are going to successfully manage the long-run transition, while lifting the living standards of the developing world.

Index

G. Palmer, *Energy in Australia*, Energy Analysis,
DOI: 10.1007/978-3-319-02940-5, © Graham Palmer 2014